A Guide to Research Methodology

An Overview of Research Problems, Tasks and Methods

A Guide to Research Methodology

An Overview of Research Problems, Tasks and Methods

Shyama Prasad Mukherjee

CRC Press
Taylor & Francis Group
Boca Raton London New York

CRC Press is an imprint of the
Taylor & Francis Group, an **informa** business

CRC Press
Taylor & Francis Group
52 Vanderbilt Avenue,
New York, NY 10017

International Standard Book Number-13: 978-0-367-25620-3 (Hardback)

Library of Congress Cataloging-in-Publication Data
LoC Data here

Visit the Taylor & Francis Web site at
www.taylorandfrancis.com

and the CRC Press Web site at
www.crcpress.com

Contents

Preface

Recent times have seen an accelerated pace of research by individuals and institutions in search of new knowledge in an expanding horizon of phenomena. Also gaining ground are new and novel applications of newfound knowledge that could improve the lot of humanity or could pose threats of disruption, disarray and destruction. We have a wide diversity in objectives and types of research and an equally wide diversity in methods, techniques and tools used by research workers This should be clarified that by research workers we mean young academics who are pursuing their doctoral programmes, scientists working in research laboratories including those who do not otherwise qualify for research degrees or already possess such degrees, as also senior academics who advise and guide research workers. On the one hand, this diversity is an incentive for research workers to experience otherwise unknown methods and models as well as unheard-of research findings. On the other hand, this diversity may introduce an element of non-comparability of findings on the same subject matter arising from different research efforts.

The concept of Research Methodology as a subject of study by potential seekers of research degrees has been a relatively recent one. While research workers, in general, including those who may not seek a degree or have already earned one or even those who act as guides or advisors of research workers, may not always follow a generic guideline for their research activities in different specific disciplines, it is now being realized that a broad understanding of Research Methodology as a flexible framework for the research process may be quite helpful.

The present book in eleven chapters attempts to provide readers with a broad framework for research in any field. Of course, a bias toward quantitative methods and particularly toward Statistics and Operations Research could not be avoided. Of course, attention has been given to provide illustrations from different disciplines. Going ahead of common considerations in Research Design, problems of data collection using survey sampling and design of experiments as well as methods of data analysis and the associated use of models have been discussed, though concisely in the belief that readers can easily access details about these, if interested.

Chapters 1 to 4 and also the last chapter have generic contents and are meant to have general appeal. It is expected that research workers in general, irrespective of the nature and objectives of research as well as of the knowledge environment of the workers, will have to decide on certain issues in common and will find the contents of these chapters useful in resolving such issues. It must be admitted, however, that discussions on these common issues have been largely quantitative in character and will appeal mostly to readers with the requisite background.

Somewhat similar are the contents of two relatively long chapters, viz. Chapter 7 dealing with models and their applications and Chapter 8 devoted to data analysis. As will be appreciated by all, models play a vital role in any research study. Issues pertaining to model selection, model testing and validation, and model solving should engage the attention of every research worker. The amazing variety of models, the diverse fields of enquiry where these can and should be applied to reveal special features of underlying phenomena, the increasing diversity of ways and means to solve these models including methods for estimating parameter models and, finally, a recognition of model uncertainty are topics which have occupied tons of published materials. In fact, many of these topics have been discussed in complete books devoted separately to each of them.

The chapter on data analysis attempts to briefly highlight several different problems in data analysis some of which may be encountered by any research worker. The discussions contained are mostly illustrated in terms of numerical examples drawn from as wide an ambit as possible. The discussions are not meant to explain even briefly certain methods which are known to be useful in analysing some types of data. Rather, some types of research problems and correspondingly relevant data which may involve several different methods or techniques to analyse have been taken up. In some cases, limitations of existing methods for data analysis have also been pointed out.

Two important methods for generation of evidence for theory-building or theory-testing exercises, viz. sample surveys and designed experiments, are discussed in Chapters 5 and 6. The intention was not to consider the usual procedures and results in connection with these two methods as they are explained in well-written text and reference books.

Not much originality is being claimed for the content and presentation of multivariate data analysis dealt with in Chapter 9 and of analysis of dynamic data in Chapter 10. In fact, most of the more important topics and sub-topics in the area of time-series analysis have been left out in the hope that the research workers in need of these methods will access relevant publications to meet their needs and even to know more about these problems and their existing treatments. In contrast, certain apparently unimportant issues in these contexts which do not find place in erudite writing but are matters of concern in real-life investigations have attracted the attention of this author and found some delineation here.

The content of Chapter 11 lacks in details and is somewhat sketchy. One reason for this the author's conviction that not all such details can be worked out and put in print unequivocally.

In terms of its intended coverage, this book is meant not merely for empirical research in the perceptual world involving imaginative applications of known methods and techniques but also to help those who are or will be engaged in theoretical research even in the conceptual world. One distinct feature of this volume which should justify the appreciation of inquisitive minds is that it explores research problems in uncharted fields. Not unduly claiming to possess knowledge in all such fields, the author tried his hand in interacting with scholars from a wide range of interests to get an idea of problems which await complete or even partial solutions.

The author has tried to delineate the need for further research while discussing subjects that have grown sufficiently over the years and have been adequately covered in text and reference books.

With its content largely focused on materials not covered by the usual textbooks, the present volume is primarily addressed to advanced students who have undergone some expository courses in quantitative methods.

It has been an adventure to write a book on Research Methodology that would claim at least some amount of originality in content, coverage and presentation. An equally important consideration is the need for a 'balanced' presentation on different topics. While the objective of putting in some original and some recent materials was sincerely attempted, the second objective of striking a balance in the depth and breadth of discussion and the inclusion or absence of illustrations in respect of different topics will not evade the discerning eyes of a careful reader.

Improvements will be welcome to make the volume more useful and the author keenly looks forward to receiving suggestions for such improvement from all interested quarters.

S.P. Mukherjee

Acknowledgements

I first felt the urge to prepare some material for the benefit of research students and professionals engaged in research as individuals or as members of research teams nearly a decade back when I was requested to conduct a short course on Research Methodology for research students in the Department of Statistics, University of Calcutta. I also accepted invitations to teach the course in several other institutions. I greatly appreciated invitations to address research students on some specific topics. In fact, I greatly enjoyed such an invitation from the Bocconi University in Milan, Italy, to conduct a short course on Characterizations of Probability Distributions for a small group of doctoral students. Similar experiences in the universities in Luxemburg, Dublin, Tianjin and a few others kindled my interest to know more about problems which I was having in my mind and which seemed to me as beyond my capability to pursue.

I enjoyed helping many professionals in the fields of Management, Engineering and Medicine in planning their empirical researches and in analysing the data they compiled to make inferences about diverse phenomena of interest to them. This activity has kept me engaged over the past five decades and has widened the ken of my vision.

I felt the need to learn more and to think somewhat generically about the tasks any research worker has to face. I could gradually distinguish in my own way between Research Methodology and a collection of Research Methods applicable in different situations involving analysis and interpretation of data. Subsequently, I plunged into the task of writing a book on the subject that could possibly convey this perceived distinction. At the same time I doubted my own ability to focus adequately on Quantitative Methods commonly needed by research workers drawn from various disciplines. And I did realize that most research would involve both qualitative and quantitative analysis.

I remain grateful to the Statistics faculty in the University of Calcutta (most of whom happen to be my former students) for whatever has eventually come out in this volume. The opportunity I received to conduct the short course for research students and to interact with them on various aspects of their research activities helped me immensely. I should also thank many academics and scholars among my former students as also my acquaintances in various disciplines for the variety of problems that I could somehow discuss in the present volume.

Sincere thanks are not formally needed and are not enough to put on record the constant encouragement and profound support I received from my wife Reba. I had many useful discussions with her about research in general and research in behavioural sciences in particular. My sons Chandrajit and Indrajit were a source of great inspiration. They would enquire regularly about the progress at my end during the overseas calls they made every day.

I will be grossly failing in my duty if I do not put here my sincere appreciation of the help and support I received from Aastha Sharma and Shikha Garg from the publisher's end during the entire journey from preparing a draft proposal to finalizing the Title of this volume and the presentation of the material contents. I must appreciate the support and guidance extended by Angela Butterworth and Roger Borthwick in preparing the final text.

S.P. Mukherjee
April 2019

About the Author

S.P. Mukherjee retired as Centenary Professor of Statistics, University of Calcutta, having been involved in teaching, research and promotional work in the areas of Statistics and Operational Research for 40 years. He guided 24 scholars for their PhD degrees in Statistics, Mathematics, Engineering and Management. He has nearly 80 research papers to his credit. He received the Eminent Teacher Award from Calcutta University (2006), the P.C. Mahalanobis Birth Centenary Award from the Indian Science Congress Association (2000) and the Sukhatme Memorial Award for Senior Statisticians from the Government of India (2013). A former Vice-President of the International Federation of Operational Research Societies (IFORS), he is currently Chairman of the Board of Directors, International Statistics Education Centre, ISI, Calcutta.

1

Research – Objectives and Process

1.1 Introduction

The word 'research' is not expected to possess a unique or unanimously agreed definition. A curious worker can find out a whole host of definitions, which are bound to cover
some common ground and yet maintain their own distinctiveness. The Merriam–Webster
Online Dictionary defines research as: "studious inquiry or examination, especially investigation or experimentation aimed at the discovery and interpretation of facts, revision of
accepted theories or laws in the light of new facts, or practical application of such new and
revised theories or laws." The definition accepted by OECD runs as "any creative systematic activity undertaken to increase the stock of knowledge, including knowledge of man,
culture and society, and the use of this knowledge to devise new applications". One can
roll out many other definitions. Etymologically speaking, the word 'research' is derived
from the French word 'recherche' which means 'to go about seeking'. Thus seeking knowledge is at the root of 'research'. And the OECD definition rightly emphasizes the creative
and the systematic aspects of this 'seeking'.

It may not be out of place to start with some largely agreed definition of Research, though
it is easier to describe what a researcher does or has done than to define 'Research'. The definition that will not attract much of a criticism states that *Research is search for new knowledge
or for new applications of existing knowledge or for both*. Knowledge, here, connotes knowledge
about different phenomena which take place (often repeatedly) in the environment or the
economy or the society – together constituting the perceptual world – or even in the conceptual world of human imagination or thought. To elaborate, knowledge encompasses
Concepts, Models, Methods, Techniques, Algorithms, Results and Software etc.

The ultimate output of Research is to explore new knowledge by processing various relevant and inter-related pieces of information generated through observations or experiments
or experiences. Such information items have to be reliable and reproducible (from one
location or institution to another) and the knowledge they generate has to be accessible
by all interested parties. A major distinction between Research and Development could be
the relatively larger uncertainty characteristic in a search for new knowledge, compared
to that in an attempt to apply that knowledge and come up with a concrete entity in a
Development activity. At the same time, both during research into a phenomenon and
in development of a concrete entity by suitably applying some newly found or existing
knowledge, we are quite likely to come across steps and consequences which could have
been better avoided since those were not putting us on the right track. However, such
avoidable steps and undesirable consequences were only to be found out, since those were

not just obvious. And such findings constitute valuable information that can be used to develop better procedures or algorithms and should not be dubbed as simply 'negative' and a wastage of efforts. In fact, such information is as important a research output as a 'positive' one. This is applicable in an equal measure to activities to make use of existing knowledge to come up with some concrete entity.

We generally restrict ourselves to Research activities in the perceptual world and, there again, to research areas where some outputs are targeted to benefit not merely academics but also a large cross-section of the people in terms of coming up with a process or a product or a service or a method or a practice or even a concrete and comprehensive idea that could lead to any of these entities. However, researches can be and are taken up in the conceptual world as well. In fact, in Mathematics and in Mathematical Sciences research primarily concerns entities which are not concrete but abstract even when those relate to the perceptual world or concern phenomena which remain in the conceptual world, e.g. activities of the human mind.

Research findings – specially those of seminal or basic or significant researches – result in discoveries and inventions. In the perceptual world, we can have both discoveries and inventions. We discover new materials, structures, functions, interactions along with uses and abuses of concrete entities which were existing already, but beyond our knowledge and now found out. Research in this context consists in 'finding out'. We also invent new entities – abstract as well as concrete – which did not exist earlier. The steam engine best illustrates invention of a concrete entity. Several methods for numerically solving algebraic or differential equations are inventions of abstract entities.

Most research workers are engaged in attempts to discover some aspects or features or functions (of existing entities) hitherto unknown.

To talk about research in the conceptual world, we enter the Theory of Numbers and come across many innocent-looking conjectures relating to properties possessed by numbers, lying beyond our knowledge and awaiting complete proofs. With such proofs, we would end up with interesting, if not important right now, discoveries. Thus, Catalan's conjecture put forth in a slightly different form by Levy Ben Garson in 1342 that among all possible pairs of integers which can be expressed as powers of whole umbers, only 8 and 9 are consecutive numbers, proved in 2004 by Preda Mihailescu, is a genuine discovery.

It must be admitted that not all research works end up either in a discovery or an invention. Many empirical research works may sieve out new knowledge (confirming or refuting existing knowledge or belief or action) from strings of information on several issues related to the phenomenon under investigation that are extracted from data (observations without intervention and/or involving some intervention). And such new findings, particularly those relating to social and economic phenomena or systems, may be quite useful for policy formulation and evaluation.

In simpler terms and to focus on empirical research, some authors look upon research as an inquiry into the nature of, reasons for and consequences of a particular set of circumstances, whether these circumstances are experimentally controlled or just observed as they are. And the research worker is not interested in just the particular results available to him. He is interested in the repeatability of the results under similar circumstances and reproducibility of the results in more general (and, maybe, in more complicated) circumstances. This way the results can be validly generalized.

It may be worth while to point out some distinctive features of social science research which has been engaging an increasing number of research workers to work on some contemporary problems arising in the society or the economy or the polity. To a large extent, social science research involves human beings and their groups and associated abstract entities like perception, attitude, personality, analytical skills, methods of teaching, evaluation of performance, merger of cultures, convergence of opinions, extinction or near-extinction of a tribe, and the like. On the other hand, research in 'hard sciences' like Physics or Chemistry or Biotechnology etc. involves largely concrete entities and their observable or measurable features.

Social Science is a big domain that encompasses Psychology, Social Anthropology, Education, Political Science, Economics and related subjects which have a bearing on societal issues and concerns. Currently topics like Corporate Social Responsibility (CSR), Knowledge Management, Management of talented or of gifted students, leadership and emotional intelligence have gained a lot of importance and have attracted quite a few research workers. While we have 'special' schools for the mentally challenged children, we do not have mechanisms to properly handle gifted or talented children.

While encompassing Economics and Political Science within its ambit, Social Science research today challenges many common assumptions in economic theory or political dogmas or principles. Some recent research is focused on the extent of altruism – as opposed to selfish motives – among various groups of individuals.

There has been a growing tendency on the part of social science researchers to quantify various concepts and constructs and to subsequently apply methods and tools for quantitative analysis of evidences gathered to throw light on the phenomena being investigated. While this tendency should not be discouraged or curbed, it needs to be pointed out that in many situations such a quantification cannot be done uniquely and differences in findings by different investigators based on the same set of basic evidences may lead to completely unwarranted confusion.

Most of social science research is empirical in nature and, that way, based on evidences available to research workers. And even when such evidences are culled from reports or other publications on the research theme, some evidences by way of pertinent data throwing light on the underlying phenomena are generally involved. And the quality of such evidences does influence the quality of inferences derived from the evidences. In fact, evolution of some special statistical tools and even concepts was motivated in the context of data collection and analysis in social science research. While the dichotomy of research as being qualitative and quantitative is somewhat outmoded, it is generally accepted that any research will involve both induction from factual evidence and deduction of general principles underlying different phenomena and is quite likely to involve both quantitative analysis and qualitative reasoning. In fact, uncertainty being the basic feature of facts and factual evidences about social phenomena, we have to use probabilistic models and statistical tools to make inductive inferences. It is this recognition that can explain two generic observations. The first is that quite a few statistical concepts, methods and techniques owe their origin to problems which were faced by research workers investigating individual and collective behaviour of human behaviour in different spheres of their activities and the impact of the latter on the economy, the society and the environment. The second relates to the fact social science research has not always taken full advantage of emerging concepts, methods and tools in statistics to enhance the substantive – and not just technical – content of research and the findings thereof.

There are phenomena which have attracted research workers from several completely different fields. One such could be the phenomenon of 'collaboration' and collaborative work. Research in Information Systems attempts to find out how the collaborative efforts of different groups of people result in software development. Historians carry out research to understand how cathedral builders managed to raise a multitude of tall stone cathedrals all across Europe in a relatively short period of time. The ethnographer grapples with the fact that field service technicians collaborate on fixing broken copying machines. While collaboration is a phenomenon that one comes across in many situations, it becomes a matter of research when the result of collaborative work is something unusual and needs explanation.

Fundamental research likely to continue with answers to some basic philosophical issues based on experiments and experiences as these accumulate can be illustrated by the question: how does a community of practices differ from a practice of communities? The first can be built around some theory, while the latter is woven around some widely adopted practice. Is 'theory' divorced from theory or, at least, can theory be cultivated regardless of 'theory'? Further, can a community of practices grow up without a supporting theory? How real is the theory–practice dichotomy?

Incidentally, one is reminded of the famous remark by Zimmermann that "No Pure mathematics is good unless it finds a good application: no Applied mathematics is good unless it is based on sound theory". But then the issue has not been settled forever and research may continue. Similar is the case with the matter–mind dichotomy, which is just wished away by the great neuro-photo-biologist Adelman, who in his book with the fantastic title "Bright Air, Brilliant Fire" makes an initial remark: "What's matter? Never mind. What's mind? Doesn't matter."

1.2 Research Objectives

Research findings in the perceptual world add to the existing stock of knowledge regarding different processes and resulting phenomena which take place in the universe. Such processes and consequential phenomena are usually repetitive in nature and are governed by some law(s) Thus we have the gas laws $PV = RT$ or $PV^\lambda = k$ (a constant) governing the phenomena arising from subjecting an enclosed mass of gas – monatomic or otherwise – to changes in pressure (P) and temperature (T) resulting in changes in volume (V). We have Fechner's law $R = c \ln S$ governing the relation between the degree or extent of stimulus (S) applied to a living being and the response (R) (of a given type) obtained. Laws need not be quantitative in character. And not all laws of a quantitative nature can be expressed by way of equations. In fact, an important aim of research is to identify the form(s) of the law(s), estimate the parameter(s) involved like R or λ or c and even verify such laws whenever needed, on the basis of observations on the underlying processes or phenomena.

The objective(s) of any research, to be reflected and broadly indicated in the Expected Outcome or Working Hypothesis(ses), could be

1. Creating or developing or building up some new concepts or methods or properties/ features or results or products like software and the like or a combination of some or all of these. Speaking of research in Statistics, concepts of copulas to represent inter-dependence among random variables, concepts of relative entropy or distance between probability distributions, concepts of concentration like the Simpson or Zenga index, concept of time rate of unemployment or under-employment etc. We can think of methods of sampling like chain referral sampling or the use of rotational panel surveys or methods for constructing designs for industrial experiments which can satisfy some real-life constraints like incompatibility of some treatment combinations in the context of factorial experiments or the probability weighted method

of moments for estimating parameters in some probability distributions and similar other methods and results to illustrate objectives of further researches in the field. In the field of Medicine, the objective could be to develop a new and maybe simpler and/or cheaper non-invasive method to detect a difficult disease, or to conduct clinical trials with some new medicines or treatment protocols to compare their relative efficacies or to find out the pathway of action for a medicine or even to examine the possible role of robots in carrying out difficult surgeries, and so on. In recent times, we come across functions of several variables (some or all of which could be integers) which are highly non-linear, are multi-modal with several local maxima (minima) to be maximized (minimized) subject to some constraints. Many of these optimization problems await optimal solutions through known algorithms which converge to a solution which can be algebraically proved to the global optimum. In such cases, even a near-optimal solution that can be worked out by using a meta-heuristic search may serve the purpose. And research has been going on to work out acceptable near-optimal solutions through finite-time computation.

2. Generalizing some existing models or methods or techniques to widen their applicability, we find umpteen examples in the area of Probability and Statistics, e.g. generalizing the exponential model to the Weibull or the gamma or the generalized exponential model, generalizing the logistic law of growth to include time-varying rate of relative growth or to absorb a constant to ensure a better fir to observed population figures in different countries and communities, generalizing the birth-death process to the birth-death-immigration process, etc. We can also consider generalizing the concept of regression based on conditional expectation to represent dependence of one variable on another by taking into account certain properties of the joint distribution along with the marginal distributions, as also properties to summarize the conditional distribution other than its expectation. We thus have concepts of positive quadrant dependence or stochastically increasing conditional distribution or right-tail increasing property and the like.

    Compound distributions to cater for random environments or mixtures – finite or using a continuous mixing distribution; from entropy to entropy in past or in residual life or in record values also illustrate generalizations which the existing entities as particular cases.

3. Extending to wider or more general set-ups, e.g. extending some probability distribution from univariate to multivariate situations; extending a method of inferencing from the classic 'independently and identically distributed observations' to dependent processes and the like. It must be kept in mind that not too seldom such extensions may not be unique and sometimes multivariate extensions of univariate distributions may not behave in tune with the corresponding univariate distributions. An interesting example is the fact the loss-of-memory property of the exponential distribution gives rise to four classes of bivariate exponential distributions having. very weak, weak, strong and very strong versions of the loss-of-memory property. It may be mentioned that most bivariate exponential distributions have the very weak loss-of-memory property and possibly none is known to possess the strong version of this property.

4. Modifying some existing models or methods or results to remove certain constraints or to take them into account explicitly. Research questions have been raised and more are likely to be raised to relax the assumption of integer order of moments in estimating distribution parameters, or to drop the property of symmetry from the

normal distribution retaining some of its other useful properties. Similarly, we may think of recognizing explicitly certain constraints like the mean strength exceeding mean stress in estimating reliability or some inequality relations among regression coefficients in estimating a multiple linear regression equation. Illustrations can be easily multiplied.

Differentiation among these objectives, particularly among the last three, may be pretty difficult and, possibly, redundant. They can be easily grouped under a broad canopy.

The above statements relate mostly to researches in Statistics and its applications. However, similar remarks apply to research in general, though exploratory research into a completely new phenomenon or system may have a limited objective to provide a description of the system or phenomenon and relations among the different descriptors or to offer a provisional explanation of some aspect of the phenomenon or system studied while confirmatory research usually attempts to confirm or discard some such explanation for a system or phenomenon investigated earlier. The next section deals partly with differentiation among researches of different types in terms of research objectives.

1.3 Types of Research

Depending on the types of processes and resulting phenomena being studied or the objectives and expected outcomes or the design adopted or the way research was initiated or mooted, the manner in which resources were mobilized or the scale of generality of the research findings or the type and depth of analysis or similar other considerations, research can be put into several categories, essentially to reflect the specific features of a particular research study. Such classifications are not all orthogonal and one may not be able to have an exhaustive classification that can accommodate any research work already done or to be completed in future.

Thus, one may think of theoretical and empirical research. And theoretical research is not necessarily confined to the conceptual world. It can very well investigate some processes or phenomena in the perceptual world, proceed on the postulate that the processes and the phenomena are governed by some laws and attempt to identify such laws on the basis of repeated observations on the processes and the phenomena. If the laws are already given in terms of findings of some previous research work, the current research study may use these observations to confirm or modify these laws or even to reject the known laws. And right here we can classify research as exploratory or confirmatory, depending on the objectives. Dealing with some aspects of processes and phenomena not examined earlier, research may be directed to explore possible relations bearing on the processes or the phenomena which may subsequently lead to formulation of the laws governing the phenomena.

Industrial research or research sponsored by industry may be distinguished from other types of research in terms of a specified objective, namely to come up with a new material or product or process or service or equipment or control mechanism which will be cheaper and/or more user-friendly and/or less harmful for the environment and/or easier to maintain and/or easier to dispose of at the end of its useful life. Such a research study may well follow a proven technology and may work out necessary augmentation, correction or modification at stages wherever necessary. New product development in that way is a big area for industrial research. These may be carried out within R&D set-ups within industries or outsourced to established and competent institutions and laboratories with adequate financial and organizational support.

Some research studies are meant to throw some light on a process and the resulting phenomenon at a local level, validity being claimed for a given context delineated by a given region or a given time period or a particular population group or a specific environment (background) and the like. Others could be designed to offer more generally and widely applicable results. Taking into consideration the type of phenomena being probed and the nature of observations that may arise, as also the expected outcome(s), some research may involve a lot of quantitative analysis, while a fundamentally qualitative analysis may be appropriate in some other researches. It is, of course, true that quantification of concepts and measures has been rapidly increasing in all types of research and any research today involves both quantitative and qualitative analysis.

Sometimes, narrowly specified researches distinguished in terms of objectives and methods of data collection and of analysis are recognized within the same broad area of research. Thus psychographic or demographic or motivational research on consumer behaviour are recognized as distinct types of research within market research as a broad area. Psychographic research (also known as lifestyle analysis) is focused on activities usually carried out by the respondents (often the potential or existing consumers), their interests in different characteristics of any product – some interested in appearance, some in durability, some in ease of maintenance, some in cost price and the like – and their opinions on different products and services, on advertisements and sales promotion mechanisms, and the like. Psychographic profiles are complementary to demographic profiles. They reveal the 'inner consumers' – what customers are feeling and what is to be stressed by the firm's promotional campaign. Demographic research attempts to bring out the role of socio-economic factors besides age, gender, type of residence, household composition, educational level and similar other factors in buying behaviour.

Psychographic inventory is a battery of statements designed to capture relevant aspects of a consumer's personality (classified as compliant, aggressive or detached), buying motives, interests, attitudes, beliefs and values. Results help marketers in their search for the location of their target markets. Motivational research as the very name implies considers different means and mechanisms including advertisements, focus-group discussions, sponsorship of public events etc. to orient consumer behaviour towards some specific brands or grades. And market research which is imbedded in research on consumer behaviour is itself an important component of research on human behaviour which pervades the entire gamut of social science research.

A somewhat different type of research which does not beget the same type of recognition by the academic community but is quite useful in terms of its output is action research. In fact, action research is meant to solve problems or to effect improvement in systems or processes by investigating different possible interventions in the systems or manipulation of the variables in a process, getting those implemented by involving the potential beneficiaries and evaluating their effectiveness to eventually recommending a policy or an action. Such research is quite common in social sciences and their outputs benefit both the researcher and also the group facing the problem. In the broad sense, action research identifies some problem being faced by a group or in an area – which could be economic or social or political – and investigates the problem closely, throws up alternative interventions or actions in the underlying system or on the process concerned, analyses data on the relative merits and limitations of the different alternatives to find out the most effective (of course feasible) intervention or action to be implemented in order to solve the problem.

1.4 Research Process and Research Output

Research can be comprehended both as a process as well as the output or outcome of such a process. Although research activities in different subject areas and taken up by different individuals or teams under different conditions are quite different among themselves, we could draw upon the basic features common to all of them. Thus, for example, we agree that research is a process with many inputs of which knowledge in the relevant domain along with a spirit of enquiry is possibly the most important and that the output or outcome of research is something not very concrete always.

In recent times when researches are taken up more by teams rather than single individuals, more in terms of costly and sophisticated equipment and facilities being deployed rather than using make-shift devices, more thorough long-drawn-out experiments than by using a few observations or trials, the research process has to be well-planned.

Let us look at the broad steps in research, covered in the following list offered by many writers on the subject.

a) Concrete formulation of the Research Problem, indicating the processes and/or the resulting phenomena to be studied along with their aspects and concerns as noted in the problem. As research progresses, more dimensions/aspects relating to even some more processes and phenomena than initially envisaged may also be covered. The focus is on clarity in the initial formulation in a manner that can admit of subsequent expansion.

b) Review of relevant literature in the Problem Area identified by the research worker and available experiences, if any. This does not imply just a summarization of the methods, data and findings as reported in different relevant articles and reports already published or otherwise accessible. What is more important is an assessment of gaps or limitations in the literature reviewed in respect of models used, data analysed and methods of data analysis, interpretation of results obtained and the like. Thus the review has to be a critical one that can suggest problems to be taken up by the reviewer.

   [These two steps may not follow this sequence and the first step may only be possible after scanning the extant literature in the problem area.]

c) Formulation of the Expected Outcome or the Working Hypothesis in terms of removal of concerns about the problem or of limitations in earlier investigations and results thereof. In the case of a research problem not studied earlier, one has to delineate the depth and breadth of the expected outcome.

d) Research Design which provides the essential framework for research and guides all subsequent steps to achieve the research objectives as are reflected in the Problem Formulation and the Working Hypothesis. The Design delineates the nature and amount of data to be collected to achieve the research objective(s) and/or to demonstrate that the objective(s) could be achieved, the way the data have to be collected and checked for the requisite quality, the types of analysis to be carried out, the manner in which results of analysis have to be interpreted in the context of the research objective(s) as also possible limitations of the research findings. The role of Research Design goes even further to include procedures to establish the validity of the research findings.

The overall research design in empirical research involves in the widest sense (1) the sampling design if we need survey data, (2) the experimental design if we have to conduct an experiment to yield responses which will provide the required data and (3) the measurement design to tell us how to measure a feature or characteristic or response corresponding to an experimental (in a laboratory or field experiment) or observational unit (in a sample survey).

The design must also identify (a) experimental units (in the case of a laboratory or field experiment) and treatments, (b) factors or independent variables and response(s) as dependent variables, (c) covariates other than the factors which may affect the response(s), (d) exogenous variables given to the system under consideration from outside and endogenous variables developed within the system in terms of causal dependence among variables within the system, (e) (unobservable) latent variables and (observed) manifest variables, (f) confounded relationships and (g) experimental and control groups.

e) Collection of evidences in terms of primary and/or secondary data throwing light on the research problem in the case of empirical research and of all supporting information, methods, tools and results in the case of theoretical research. Sample surveys and designed experiments being the two well-established methods to collect evidences, we have to develop appropriate instruments for collection of data, e.g. schedules of enquiry or open-ended questionnaires for sample surveys and mechanisms to hold factors at specified levels or values in designed experiments. It is also pertinent that while adopting a particular sampling design or a particular experimental design in a certain study we need to work out suitable modifications or extensions to suit the specific features of the given situation.

f) Analysis of evidences through use of logical reasoning, augmented by methods and techniques, both qualitative and quantitative, to come up with a solution to the problem. It must be remembered that use of more sophisticated methods or tools of analysis does not necessarily lead to a better or more efficient analysis. The type of analysis as also the tools to be used depend on the objective(s) of the analysis, the type of data to be analysed (e.g. whether missing data or outliers are present or whether the data were censored or not, etc.), the nature of models which can be justified and similar other considerations.

g) Establishing validity of the research findings to ensure face validity, concurrent validity and predictive validity. There are different forms of validity and each form needs a particular method of establishing that form of validity. Face validity can be easily checked by doing some numerical calculations and verifying if the final results are absurd or not. For empirical research, establishing predictive validity or even concurrent validity against some comparable results currently available is an important exercise. Checking whether a general form or model or method or result does include known forms or models or methods or results as particular cases is in some cases an appropriate validation exercise. In empirical research, triangulation is a method to validate results obtained by different methods used by different investigators on different groups of subjects.

h) Preparation of the research report and dissemination of research findings to all concerned before the same are distributed widely or even submitted for publication or in-house documentation. Feedback from stakeholders including peer groups in the case of theoretical research should be comprehensively analysed and necessary modifications and/or corrections should be incorporated before wider dissemination. Publication ethics should be strictly adhered to.

To distinguish Research Process from Research Output, one may possibly consider the first five steps as corresponding to the process, steps (f) and (g) to the output and the last step to communication/dissemination of the output. It may not be wrong to combine the last three steps to discuss the quality of research output, and to speak about the quality of research. We should definitely consider inputs into research and their quality.

Inputs to the research process include, among other elements, (1) motivating factors, expectations of the organization and of the research community or even the concerned segment of society, (2) documents, patent files, standards for processes and procedures likely to be followed in executing the process besides the most important soft input, viz., (3) intellectual and creative ability of the research worker(s) and (4) software for simulation, computation and control. Quality of inputs into the main process as outlined in the steps stated earlier as also of support processes like equipment, materials and utilities, work environment and the like turns out to be important in affecting the quality of the research process. And quality of some of the 'hard' inputs can be examined in terms of calibration of equipment and the resulting uncertainty in measurements to be generated, process and procedure standards being up-to-date, available software having requisite efficiency, laboratories having control over ambient conditions, reference materials being duly certified, etc. Knowledge of the subject domain, of relevant models and methods, of algorithms and software and the like can also be assessed in broad categories if not in terms of exact numbers.

In research, conversion of the so-called inputs into what ultimately will be treated as the output is so complicated and subject to so much uncertainty that relating quality of output to quality of inputs in an acceptable form may be ruled out. There could be cases where some relations – if not quantitative – can be established and made use of to ensure quality in all conceivable inputs.

Research, in general, may not even allow a formulation of its scope and objective(s) right at the beginning and may have to grope in the dark in many cases. The problem area may be, of course, identified as one that is currently associated with some doubt or confusion or ignorance or uncertainty and may even be one which has been investigated by earlier research workers. The only delineation of the Research Problem in such cases is provided by an outline of the remaining doubt or confusion or ignorance.

In this sequence of steps, research design is the most crucial and plays the role of a process plan in the context of a manufacturing or service industry. And quality in the research process combines the concepts of quality of design and quality of conformance (to design requirements). In modern quality management, quality of design outweighs quality of conformance and this is verily true of the research process.

Incidentally, there are strong advocates of complete flexibility in the Research Process with no initial directions or guidelines. To them, the Research Design cannot be pre-specified at the beginning: it evolves gradually as the investigation proceeds, with steps at the next stage not known until some implicit or explicit outcome of the earlier stage is noted. Using the relevant language, the Research Design is sequential and, even if pre-specified, calls for modifications and even reversions as and when needed. However, this flexibility may be desired in research meant just to acquire new knowledge in the conceptual world more in the case of sponsored research or new knowledge in the conceptual world than in the case of sponsored research or applied research in the perceptual world.

Action Research, particularly in the context of some educational or socio-economic issues, which is meant to improve some current processes or systems, e.g. of teaching History to secondary school students or providing some healthcare to people who are handicapped in some way or another, usually involves the following steps:

Diagnose the pre-intervention situation and specially the problems faced therein. This may require planning and conduct of a base-line survey.

Analyse the survey data and the information collected from other sources.

Identify the key issues relating to the problems at hand and develop an action plan to resolve these issues.

Check the feasibility of the action plan, by considering possible co-operation and/or resistance from the people affected by the problems.

Modify the action plan, if needed, to make it feasible and launch the programme of implementing the planned intervention or action.

Monitor the implementation at each stage and carry out concurrent evaluation of the effectiveness of the plan to remove or reduce or lead to a resolution of the problems

Once implementation is completed, evaluate the outcome of the planned action and its impact on the potential beneficiaries. If found successful, make suitable recommendations or suggestions for replication of the action plan for similar purposes in other areas or for other affected groups. Otherwise, analyse reasons for failure by using techniques like brainstorming followed by Ishikawa diagrams and Failure Mode and Effect Analysis. Try out the modified action plan maybe after a reasonable time gap in the same area involving the same group of affected people or in a different area with some features shared by the first area.

1.5 Phases in Research

The steps in the research process outlined in the previous section are quite generic in nature and are involved more or less explicitly in all research studies. However, these steps corroborate what may be called a neutralist approach to any enquiry with the researcher as an individual with his/her personal background covering knowledge, skill and 'academic politics' besides personal biases or preferences and priorities having no role in the process, except when it comes to interpreting the results obtained. In researches dealing with living beings – their behaviours and experiences – it may be argued that such a neutral stand may not be and even should not be insisted upon. In fact, models for social science research stresses the 'philosophical' aspects as also the aspect linked with the 'context' in the entire research process.

Knight (2008) refers to four phases in the research process, viz. conceptual, philosophical, implementation and evaluation. The conceptual phase includes: (a) the research: the single phenomenon or the group of phenomena with possible inter-relationships to be studied, the research topic and the specific research question(s). Also to be stated here is what knowledge about which aspect of the underlying phenomenon or phenomena is likely to be achieved; (b) the research discipline: to indicate the discipline(s) from which concepts, methods and techniques will be used and to which some knowledge augmentation is expected to take place; and (c) the researcher: taking account of the knowledge and skills of the researcher and the researcher's 'theoretical lens', reflecting the ability of the researcher to develop a preliminary or working hypothesis(ses) which can lead to the formulation of new theories or modification or augmentation upon verification of existing theories. These three elements correspond to the conceptual phase and define the 'point of view' that determines the succeeding phases. To the **positivist** researcher, processes of data collection, analysis and interpretation should not recognize the role of the researcher. On the other hand, the researcher is part of the world being studied and thus the researcher's own influencing point of view is inevitable to the critical and **interpretivist** approaches.

The point of view determines the philosophical stage and the research epistemology through the assumptions being made. To the **positivist** relevant questions are "is the world objective? measurable? independent of the research instruments?" Corresponding questions to an **interpretivist** are "is the world observable? Can it be influenced? is it relational?" The first approach begins with a hypothesis which the researcher proceeds to test by adopting the relevant methodology. In the second approach, research begins with observing and analysing the phenomenon of interest and develops possible explanations regarding its characteristics of interest. In the words of Amaratunga (2002) "Epistemological pluralism endorses both quantitative / deductive and qualitative / inductive approaches by supposing that both approaches have degrees of the other within them".

The third or implementation phase involves the research design, which is the most engaging component of research methodology as a framework for the research enquiry. Here we have to identify efficient methods for collection of data that are necessary and sufficient for the purpose as also the research methods for data analysis. Data analysis may have to serve different objectives, viz. exploring relations, confirming or refuting existing hypotheses, proposing explanations by way of hypotheses being developed in the research, predicting outcomes of current happenings or decisions, etc.

The last phase concerns evaluation of research results in terms of validity and generalizability. It is also possible to evaluate a context-specific research whose findings are not meant to apply in other contexts but should be evaluated in terms of internal consistency. It is quite important for the researcher to note down limitations encountered during different steps in the research process, right from attaching suitable operational definitions to the variables of interest to problems of securing informed wilful responses or measuring the value of the response or yield from each experimental unit, matching the data obtained with the objectives of research in terms of relevance, adequacy and accuracy., accepting as valid assumptions implied by some tools for analysis, difficulty in obtaining even a satisfactory solution to a complex computational problem, and many similar issues all of which eventually impose limitations on the validity and, beyond that, utility of the research outcome. It is quite likely that the researcher will note down and admit such limitations, without being able to remove any of those within the present research work. Hopefully, such limitations will be appreciated by research workers to join the quest for knowledge in this topic subsequently and offer some solutions to some of these problems arising in the conceptual world along with some arising in the empirical world as well. As remarked earlier, these four phases may not be identifiable explicitly in all research studies, but are involved with or without explicit mentions, with some phases stressed more than certain others, and some phases found missing on the face of it.

1.6 Innovation and Research

Innovation is the application of new solutions (based on science and technology and found out by the innovator) that meet new (felt, imagined or even anticipated) requirements or unarticulated needs or even existing (but unmet) needs, through new products, processes, services, technologies or ideas that are readily available to markets, governments and society. Obviously, innovation involves a new application – not made earlier – and hence can qualify to be called 'research'. However, innovation may not necessarily go through a comprehensive research process and the outcome may sometimes appear to be somewhat simplistic. What really distinguishes innovation from a research exercise is the nature and use of the output. In an innovation exercise, the output has to be something concrete that penetrates the market, reaches the people at large and contributes to the national economy. In research, generally, the output may or may not be concrete. In fact outputs of high-level research could often be some general principle governing some phenomenon. For innovation, an effort or entrepreneurship to convert some research output to a concrete entity is a must.

We have innovative products and services resulting from imaginative applications of new knowledge gained through contemporary scientific and technological research or even researches carried out quite earlier with findings duly disseminated but not applied by anyone this far to come up with a new, novel and non-trivial entity with a distinct market value. In fact, knowledge gained through scientific research may have to be augmented by some new technological know-how to be developed to yield something concrete with a lot of use value along with exchange value and even prestige value. Such innovations are really noteworthy and have impacted human life in significant ways. On the other hand, indigenous knowledge which might have been used in crude forms to end up in crude objects which, no doubt, were found to be quite useful, can now be used in a more imaginative way and draw upon some augmenting technology which also could be pretty simple to come up with surprisingly useful and valuable objects of innovation.

In terms of their benign or harmful impacts on human life. on environment, on production processes, on social systems, on national economy and such other consequences, we may not be able to prioritize these two broad types of innovation unequivocally. However, for individuals and institutions connected with research, the focus will remain on the first type and, in that way, on creation of new scientific knowledge, development of new technologies and creative application of these to result in meaningful innovation.

While it is true that some of the great innovations which have changed the world like that of the steam engine were worked out by creative individuals who did not follow a prescribed course of action or a 'methodology', it must also be remembered that to institutionalize innovations and to leave innovations to the exclusive fold of some creative individuals, theories have come up and have been implemented in many organizations. Notable among such theories is TRIZ (in English TIPS or Theory of Inventive Problem Solving), propounded by the Russian engineer-cum-scientist Altshuller. TRIZ recognizes contradictions as problems in technical systems – small or large, simple or complex – and provides solutions to remove such contradictions. TRIZ encompasses a 39×39 matrix of contradictions, 76 solutions and a database of solutions resulting in innovations. TRIZ has undergone some modifications subsequently, such as ARIZ, and there is a software version of the tasks to be done.

Even institutionalized innovation should proceed systematically, as has been indicated in ARIZ, developed by Altshuller. At the beginning, the intention in developing ARIZ was to invent a set of rules for solving a problem, where solving a problem means finding and resolving a technical contradiction or, for any given solution, ensuring less material, energy, space and time used. The method was intended to include typical innovation principles, such as segmentation, integration, inversion, changing the aggregate state, replacing a mechanical system with a chemical system, etc.

ARIZ 56 is a set of steps based on the practices of the best inventors. It also incorporated the idea of thinking beyond the boundaries of the immediate subject.

ARIZ 61 had three stages.

- I. Analytical Stage
- II. Operative Stage
- III. Synthetic Stage

Stage 1, the analytical stage, involves the following steps:

- Step One

State the problem. This is the most important step. The problem has to be stated without jargon.

- Step Two

Imagine the Ideal Final Result (IFR). At this stage, it is not important whether the problem can be solved as per the Ideal Final Result. An insight into IFR will give us the ultimate goal to be achieved.

- Step Three

Determine *what* interferes with achieving this result, i.e. determine the contradiction that exists.

- Step Four

Determine *why* it interferes, i.e. the reason for the contradiction.

- Step Five

Determine under what conditions it will not interfere.

The operative stage is the main problem-solving stage. The following steps will elaborate the guidelines of this stage.

Step 1: What are the possibilities of making changes in the object or system or process technology.
- a. Can the size be changed?
- b. Can the shape be changed?
- c. Can the material be changed?
- d. Can the speed be changed?

- e. Can the colour be changed?
- f. Can the frequency be changed?
- g. Can the pressure be changed?
- h. Can the temperature be changed?
- i. Can the working conditions be changed?
- j. Can the time of operation be changed?

These are general questions. They are indicative and not exhaustive. The central idea in this step is to think of what will happen, if this is changed.

Step 2: Explore the possibility of dividing an object into parts.
- a. Identify and isolate a weak part.
- b. Isolate the crucial parts which are necessary.
- c. Identify identical parts and separate them.

Step 3: Is there any possibility of altering the outside environment?
- a. Can any parameter of the environment be altered?
- b. Can an environmental parameter be replaced?
- c. Can environmental factors be separated into several media?
- d. Can any gainful function be achieved by environmental parameters?
- By this step, we are trying to exploit any environmental resources or energy that can be put to productive use.

Step 4: Is there any possibility of making changes in the neighbouring object?
- a. Examine whether there is any relation between independent objects participating in the same function.
- b. Examine whether we can eliminate one operation and transfer its function to another object.
- c. Examine if a number of objects can be operated parallelly on a defined area.
- By this step we try to assess the interaction between independent objects and try to eliminate or transfer this or to do both.

Step 5: Examine whether a similar problem has been solved in another industry, by any another technology, by anybody else, anywhere.

Step 6: Come back to the original problem. If any of the above steps could result in a solution well and good. Otherwise, redefine the problem in a more general way and re-examine

The next stage is the implementation stage, involving the following steps:

- Step 1: Try and change the shape of the object. It is normal for a new machine with a new function to have a new shape.
- Step 2: If some corresponding or interacting object or a system has to be changed or modified this has to be done.
- Step 3: Examine whether such a solution is applicable to any other object or function of the system.

ARIZ 85 – B & C

- Significant structural changes were made.
- The link between the algorithm, the system of standard solutions and patterns of technological evolution became stronger.
- Up to this time Altshuller was concentrating his efforts on ARIZ improvement. But he started ignoring suggestions saying ARIZ 85 C was adequate and started concentrating on the Theory of Development of a strong creative Personality (TRTL), ignoring ARIZ.
- As can be seen, ARIZ has been evolving and getting modified by the inputs from recent practitioners in the Western and European world.

What is important to note again is the emphasis placed on being systematic and disciplined to avoid unnecessary delays, wastes and regrets in carrying out sponsored or oriented or just applied research to transform some known results in science and technology to come up with a concrete innovation. While ARIZ and TRIZ may appear to be dealing only with 'technical systems', any system in social science even can be comprehended as a 'technical system' with relevant amplification.

Even when we consider Innovation as much less of an effort than what is called for in basic or fundamental research, along with the resulting concrete entity which is expectedly of a much lower value compared to that emerging out of some basic or fundamental research, adoption of a systematic approach or a protocol has been quite well accepted by organizations focused on innovations as a strategy to boost their performance. In fact, adoption of an approach that can be characterized in some sense as 'research methodology' has been very much appreciated in the field of innovation.

1.7 Changing Nature and Expanding Scope of Research

Mankind has been engaged in research (in the generic sense of search for new knowledge) almost since its appearance on earth. Faced with the many problems of survival and existence and subsequently for comfort and convenience, humans have been engaged in observations, in experiments, in arguments, in calculations, in validation of findings, in documentation, in interpretation and, to sum up, in research. Earlier equipments available for observations, materials and instruments for experimentation, resources required to procure experimental objects and similar other necessaries for research were quite limited. Researches were carried out mostly by individuals working by themselves alone or, at best, in collaboration with a handful of workers in the same area. Institutions exclusively meant to carry out research were countably few and research was carried out mostly in institutions of higher learning.

The scope of exploratory or confirmatory research was limited to phenomena which could be observed in nature, in society or in the economy – at least in the perceptual world delimited by phenomena or systems or processes which could be perceived by some sense organs. Thus limitations in viewing very distant or very minute objects restricted the scope of research. The non-availability of experimental materials and of materials need to study the former was another restriction.

The scope of research has opened up vigorously over the years with developments in technology equipping the inquisitive mind with accessories and instruments that enormously enhance the capacity of our sense organs to reach out to objects lying at unbelievably large distances or having unbelievably small particle sizes. Researches are no longer confined to the surface of our planet. Investigators have reached out to the crust or core of the earth, to the beds of oceans on the one hand and to outer space on the other. Emergence of many new materials and processes along with many new models and methods to make good use of such entities has changed the research scenario dramatically.

Researches in the perceptual world and even those in the conceptual world arise from keen observations (including measurements) on processes or phenomena which occur during the execution of some processes or as outcomes of these processes. Some of these processes are carried out by living beings, while others take place by themselves in nature or the economy or society. These processes are repetitive and reproducible, though in some cases like in space science a repetition of some process like the rotation of the Sun about the centre of the Milky Way galaxy may take a few million years.

Examples of such processes and associated phenomena could be:

Decay of a radioactive substance over time

Quenching of a metallic structure in brine to increase its hardness

Change in output in industry consequent upon changes in the capital and labour inputs into it

Mixing different fruit pulps or juices to prepare a ready-to-serve fruit drink to improve aroma or colour or taste etc.

Spraying molybdenum on paddy plants at appropriate stages of plant growth to enhance protein content in rice

Decrease in caries formation with increased instruction on dental care among pre-school children

Larger rate of unemployment among educated youths compared to the rate among illiterate or less educated ones

We can think of many such processes or phenomena which we come across in our everyday life or which we can think of. Natural floods, heat waves, forest fires, soil erosion, landslides, earthquakes, other natural disasters are also illustrative of repetitive processes and phenomena which we are aware of. Linked with any such process or phenomenon – natural or otherwise – problems for research are all the time being identified by some individuals or institutions.

Changing perceptions about business and its roles in society, the economy and polity along with the growing dimensions of business have motivated new approaches to business research. With customers becoming increasingly insistent on adequate quality, on-time delivery and reasonable cost, production and distribution management have to come up with solutions to problems that involve multi-person games. Risk management has become important and risk-adjusted performance measures have become a necessity for performance appraisal of executives.

Research nowadays has become more institutional than individual. In fact, many institutions have come up in subject areas with the mandate to carry out high-quality research and, in some cases, development activities to take research results to the doorsteps of households. No doubt, individuals still push the frontiers of knowledge through their research efforts. However, research teams – not always in the same labora-tory – have been making outstanding research contributions now. Most seminal research is nowadays multi-institutional. Further, research these days calls for more, better and costlier resources in terms of equipment, consumables and services than earlier, when improvisation was the order of the day. Further, modern-day research often involves expertise in many related disciplines or branches of knowledge and in that way are really multi-disciplinary. To illustrate, researches in systems biology draw upon knowledge in systems engineering and genomics along with the more recent Omics of other brands to study functional aspects of systems in living beings – from the cell to the entire organism.

Though 'research' usually calls up the picture of a 'scientific' investigation – even with a broad connotation of the qualification 'scientific' – historical researches, researches on soci-etal structures and their changes over time, researches on the evolution of different types of verses as are studied in rhetoric and prosody, numismatic researches into the changing nature of metals and materials like paper etc. used in manufacturing coins and printing notes/bills, investigations to resolve disputes about authorship of some ancient docu-ment or about the possible existence of some structures or motifs in some locations, and a whole host of similar other studies taken up in the past or being taken up to satisfy human inquisitiveness are getting more and more recognized as important research activities, not to be rated lower in worth than searches for new 'scientific' knowledge behind phenomena in the perceptual world.

In this connection, it may not be out of place to mention that any investigation that proceeds on logical lines and results in some findings which are not specific to the loca-tion and time where and when the investigation was carried out and are such that the same findings can be reached by some other investigators repeating the same investiga-tion following the same protocol as was originally adopted, can be branded as 'scientific' in a broad sense. In fact, the qualification 'scientific' has sometimes been taken to imply three different characteristics. Firstly, the investigation must be 'rational' in the sense of being related to some entity and exercise that stand up to reason and are not influenced by impulse. Secondly, the investigation should be 'repeatable', provided the phenomenon investigated is not a one-off one and it is feasible to repeat the investigation over time and/or place. The third characteristic requires the findings of a scientific study to be perfectly communicable, without any loss of rigour or value of content.

To facilitate research, research workers now have access to many sources of information, spread widely across space and over time, and make use of 'metadata' on proper checks about poolability. With advances in technology, particularly in terms of sophisticated instruments being available to measure various entities, research is now occupied more with minute variations revealed through measurements with greater inherent precision than with crude measurements or approximations

1.8 Need for Research Methodology

Do we really need some research methodology to be followed in our pursuit of new knowledge? It can be argued that significant and path-breaking researches have been done without involving a well-defined research methodology. In fact, in some such cases problems arose strangely to an individual and were solved in an equally strange or uncanny manner. We notice many remarkable developments that have changed our lives and thoughts to have taken place even before people started discussing some thing that could be christened later as 'research methodology'.

Tracing the history of science, however, one can find some sort of a scientific procedure being followed by path-breaking researchers in science and philosophy. In fact, Aristotle observed scientific enquiry as an evolution from observation to general principles and back to observation. The scientist, in his view, should induce explanatory principles from the phenomena to be explained, and then deduce statements about the phenomena from premises that include these principles. This was Aristotle's inductive-deductive procedure. A scientist progresses from factual knowledge to a reason and then deduces a general principle. In some sense, this was the basic research methodology followed in early days, without specific details about how to gather factual knowledge about an underlying phenomenon or how to come to reason with the acquired knowledge or how to deduce a general principle by reasoning.

Descartes, who invented coordinate geometry, agreed with Francis Bacon that the highest achievements of science are with the most general principles at the apex. While Bacon sought to discover general laws by progressive inductive ascent from less general relations, Descartes sought to begin at the apex and work as far downwards as possible by deductive procedure. Descartes, unlike Bacon, was committed to the Archimedean ideal of deductive hierarchy. Newton opposed the Cartesian method by affirming Aristotle's theory of scientific procedure. By insisting that scientific procedure should include both an inductive stage and a deductive one, Newton affirmed a position that had been defended by Roger Bacon in the 13th century as well as by Galileo and Francis Bacon in the early 17th century.

Given such broad directions for any scientific investigation, scientists – known as philosophers in early days – were following their own paths for observations and experiments as also for analysis and inference based on results of observations or experiments. Quite often such investigations were not focused to reach some definitive findings by way of general principles or even less general relations. There have been, however, situations where a lot of avoidable time and cost was involved while proceeding somewhat loosely in the 'hit-or-miss' approach. Trial and error have to be necessarily present in any research activity. However, adoption of a methodology is expected to minimize such avoidable losses.

Incidentally, we should appreciate at this stage the subtle but important distinction between a method or a collection of methods and a methodology – in the context of research. From what has been discussed earlier, any research involved or involves a number of tasks like planning and conducting a laboratory experiment, designing a sample survey and canvassing a questionnaire, analysis of dependence of one variable on a set of other variables, grouping similar observations into relatively homogeneous clusters or groups, comparing several alternative situations or decisions according to a given set of criteria, predicting the future state of a system from current observations on it, control of the output of a system by exercising appropriate controls of the inputs, and similar others. For each of such tasks, some methods along with associated techniques have been developed over the years and pertinent software has also been made available to facilitate computations. Depending on the type of research and its objective(s), these tasks may differ and the methods used will also differ from one research to another. A methodology, on the other hand, is more like an approach or a protocol that outlines the tasks involved, preferably according to a desired sequence and the broad nature of methods to be used and, going beyond, the manner in which results of applying the methods should be interpreted in the context of the original research objective(s). And this protocol has inherent flexibility to absorb and reflect any special features of the research under consideration. Thus a research methodology provides the superstructure, while methods are like bricks and mortar required to build up the edifice. A research methodology is like a strategy encompassing principles, processes, procedures and techniques to seek a solution to an identified research problem. In some sense, the methodology provides an architecture for the entire research exercise that determines the research methods to be applied in a given research exercise, developed to proceed from an understanding of the research question(s) and oriented towards providing direction and guidance to the whole effort to seek the answer(s) to the question(s). Research methodology provides room for creative and out-of-box thinking besides drawing upon existing research methods selectively. Research methodology pervades the entire research exercise right from data collection through data analysis and interpretation of findings to dissemination of the research output by way of documentation and publication. Research methodology has been acclaimed by some scientists as the bedrock of science.

One compelling reason for adopting a methodology is to take cognisance of recent advances in data capture, storage, transmission, analysis and interpretation. Added to this is the plethora of databases available on a host of issues which should be investigated and on which experiments are cost-prohibitive or time-consuming. There are several others. Research results have greater impacts on life and hence call for careful validation. In researches involving living beings, care has to be taken these days about ethical issues and adoption of standard protocols in this regard has become mandatory. Research designs have to be developed with a great concern for possible misinterpretation of research findings. For example, in any comparative study, the entities compared with respect to some features must be first made comparable by controlling covariates.

Before a research activity is taken up, during the research process and even after the completion of the activity, many questions do arise and do need convincing answers. And answers to such questions are addressed in the research methodology to be adopted. What follows is a set of such questions and situations which are to be considered in the research methodology.

How do we identify research problems which are to be taken up by groups supported by public or private agencies?

How can we align such problems with some socio-economic needs of the country? Or how do we address some problem areas indicated in the Science, Technology and Innovation Policy 1913 released by the Department of Science & Technology, Government of India?

We can think of developing technologies to reduce consumption of water in certain production/generation processes as also in agriculture, or cost-effective methods for arsenic removal from ground-water or for managing municipal waste water and the like. The cost of health care in the country is partly due to the absence of any design and manufacturing facility for sophisticated diagnostic and therapeutic equipment.

Experiments are almost inevitable in any research, and we know that experiments in many situations proceed sequentially. To achieve the desired degree of precision in the estimates derived from responses in an experiment, we need replications, and replications imply costs.

How many replications do we need?

Even before that, we have to decide on the factors to be varied (controlled at several levels) and the number of levels of each factor. We also have to take a stand on the number and nature of the response variable(s). What constraints are put on factors and responses by ethical issues?

How do we handle multi-response situations and locate the optimum-level combination(s) taking care of all the responses?

Results flow in sequentially and we should not wait till we have the targeted number of replications. We should look at the responses as they arrive and can even decide when to stop the experiment, as being not likely to provide the desired information or providing the same even before the planned number of replications.

Models of various kinds have to be used to analyse experimental or survey results. Once we know the mechanism of data generation and have the data at hand, how do we choose an appropriate model (from a known basket of models) or develop an appropriate model to represent the data features of interest to us?

We are often motivated to use a simulation model for a real-life process or system or phenomenon to generate relevant data for subsequent analysis and inferencing.

How do we validate a chosen model?

How do we use a model to derive the information or reach the conclusion?

How do we assess the robustness of the information or the conclusion against likely changes in the underlying model?

How do we identify different sources of uncertainty in our data? How do we assess the inferential uncertainty in the inductive inferences we make most of the time?

We should note that inductive inferences we make on the basis of evidences that bear on a phenomenon are not infallible. With adequate evidences of high quality we can provide a strong support to the conclusion we reach.

Most researches today relate to phenomena or systems which are not completely comprehended within the framework of one known branch of knowledge and call for participation by multi-disciplinary teams. Thus systems biology (which combines apparently unrelated disciplines like systems engineering with genomics), space science and technology (which require knowledge of physics, chemistry, electronics and communication engineering besides several other disciplines), environmental science and engineering (which combine chemistry, bio-technology and other disciplines) and similar other emerging areas of research do need a comprehensive research methodology to effectively engage scholars who have been accustomed to think and experiment somewhat differently among themselves in working together to achieve the research objective(s).

On many occasions, we need to generate evidences bearing on the phenomenon (an event or a relation or response to a process) being investigated – by carrying out some experiment(s) – laboratory or field – or some field surveys. Unless properly planned, we may include several factors and allow those to vary at specified levels at the cost of some of the important factors affecting the response. To study the pattern of response to change in the level or value of a factor, if we control the factor only at two levels, we will not be able to detect any possible non-linearity in response. Further, if we include even several levels of a factor which are spread only over a small range of variation in the factor, the emerging response pattern will have a restricted region of validity. In clinical trials to find out differential responses of patients suffering from the same disease to different treatment protocols like medicines, unless we follow a randomization principle or a double-blind allocation of patients to protocols, we will not be able to make valid comparisons. Similar problems may arise in the case of a sample survey as well. And response variables have to be so defined that these can be unequivocally observed (measured or counted) and so measured or observed or counted that genuine variations are not suppressed and artificial variations are not revealed.

Unless a comprehensive research methodology is developed and followed, it is quite likely that we may miss out on certain important points like checking or establishing the validity of the results obtained by using appropriate criteria and methods. We may even fail to take into consideration certain ethical issues. In the absence of such a methodology before us, we may have to wander around the wrong route to achieving our desired results. In the face of limited resources, it is imperative to make the best use of resources to satisfy the expected outcome(s). It is true that 'trial and error' is an integral part of experimentation and once we commit an error or fail to get the right answer, we are reminded of the adage "failures are the pillars of success". A methodology will help in minimizing the number of false or wrong steps and, in that way, misuse of resources.

Of the two broad categories of research, viz. break-through and incremental, even a path-breaking research which was not mandated by a policy – government or corporate – nor even taken up by the research team with the objective of solving a concretely stated research problem may not necessarily require much of resources, while a need-based research on some objective(s) to meet some national development priorities may involve a huge expenditure in terms of both manpower and physical resources. If such a research has to be wound up mid-way because of resource crunch or because the research effort was not leading to results expected at review points (toll-gates), there would be an enormous wastage of resources. As a guard against such an undesired situation, a research methodology that can detect such a possibility even at a very early stage of the research process with the advice to either abandon further work or modify the research objective(s) and/or the research design should be welcomed.

It is worth mentioning that a research methodology is not a restrictive mandate that circumscribes the imagination or the thought process of a researcher. On the other hand, adoption of some research methodology developed appropriately to suit the need of a particular research process can be a very helpful guide to avoid foreseeable pitfalls in the process to achieve the objective(s) of research more effectively. A sound exposure to research methodology followed by identification and implementation of relevant and efficient research methods and fixing the pertinent nuts and bolts is the recommended path. And mid-course correction is always a part of the entire process.

Concluding Remarks

'Research Methodology' may be a new entry in the list of phrases strewn across parts of any documented research or any dossier of research in general or in some specified field of human enquiry. However, some methodology must have been adopted in any research exercise, though not always explicitly stated in the research document. The absence of an explicit statement of the methodology followed in a research exercise is likely to result in a significant research finding not getting published or being duly circulated among peers. In fact, without a proper mention of the methodology adopted to reach a conclusion or to arrive at a new finding, any document by way of a research paper or article or note will not qualify for publication in a refereed journal of repute. Research methodology provides the approach, research methods provide the instruments – for seeking knowledge or for meaningful and new application of new or existing knowledge.

The approach to research in any area may be context-free – regardless of academic backgrounds or interests of the researcher(s) and the environment in which research will proceed and, of course, the discipline or the inter-disciplinary domain of research. However, as research becomes more organized and object-oriented, this approach becomes context-specific in most cases. And in such context-specific researches, the relevance and utility of a research methodology are likely to be and are being questioned by research workers. It may be pointed out that a systematic approach with broad guidelines about the activities or tasks to be carried out during the research study should be found useful by any investigator who can use their own discretion and wisdom to suitably amplify or abbreviate, augment or drop some of the guidelines.

The claim that outstanding and path-breaking researches have been done and documented in the past without a research methodology being a formal part of the research exercises has to be taken along with the fact that not in all such cases are we aware of the approach – not necessarily spelt out in a documented form – followed by the research workers. Curiosity to know and some trial-and-error experimentation coupled with keen observations were always the elements. Beyond these must have existed some route to arriving at the conclusions or the eventual findings that resulted from such brilliant research activities.

2

Formulation of Research Problems

2.1 Nature of Research Problems

Research problems are problems imagined or thought of in the conceptual world or faced / posed / anticipated in the perceptual world posed or formulated earlier or now, that do not have obvious solutions. Research problems may arise in both conceptual world and perceptual world. In the first, we search for new concepts, new methods, new proofs of existing results, new results or even new theory. In the second, our search relates to new properties of existing materials, new materials, design and development of new processes or products, and the like. In the first case, we come up mostly with inventions; in the second results are mostly discoveries. There are exceptions. Research problems are problems and correspond to some problem areas.

Bryman (2007) defines a research problem as a "clear expression (statement) about an area of concern, a condition to be improved upon, a difficulty to be eliminated, or a troubling question that exists in scholarly literature, in theory or within existing practice that points to a need for meaningful understanding and deliberate investigation". Among the characteristics of a research problem, mention can be made of the interest of the researcher as also of others in seeking a solution to the problem, and this interest may even amount to a concern of the interested individuals or groups about the problem and its consequences. Also important is the significance of the problem and its solution sought. It must be noted that a research problem does not state how to do something, or offer a vague or broad proposition, or even present a question relating to values and ethics.

A problem area is an area of interest to the researcher where some aspect(s) is (are) still to be studied to find a complete or even a partial solution to an existing or an anticipated unanswered question. And this is the objective of research.

Let us illustrate some problems in the conceptual world. which originate as hunches or conjectures or even from curiosity to go beyond what is already known in a problem area. Some of these conjectures or paradoxes could be solved only after centuries of dedicated efforts or have evaded any solution yet.

Problem 1. We know that $6 = 1 + 2 + 3$ and $28 = 1 + 2 + 4 + 7 + 14$. These are called perfect numbers, viz. natural numbers which can be expressed as the sums of their divisors (except themselves but including 1). The next two perfect numbers are 496 and 8218. Euler proved Euclid's observation that all even perfect numbers are of the form $2^{n-1} (2^n - 1)$ where $(2^n - 1)$ is a prime number (called a Marsenne prime). Only 49 Mersenne prime numbers have been found as of 2016 and so the number of perfect even numbers known is 49. Till today no odd perfect number has been found, though no proof exists to deny the existence of an odd perfect number. The problem is: how do we find the subsequent perfect numbers? Can we develop an algorithm for this purpose?

An attempt to resolve this problem was undertaken by Acharya Brojendra Nath Seal by defining the coefficient of a natural number as

Sum of all its divisors (except itself) / number

Unfortunately, coefficients of successive numbers do not reveal any pattern and we have no clue when this sequence of coefficients will equal 1, and this continues to be a research problem.

One may raise the question: what do we gain by solving this problem?

We should reconcile ourselves to the fact that research problems, particularly those in the conceptual world arising primarily from curiosity, are picked up and studied and also solved, if possible, not always motivated by real-life problems and a desire to solve any such problem, even remotely. However, solutions to some such problems have found extremely important applications in resolving some complicated problems of research or some real-life concrete problems, sometimes much later.

The perceptual world, which is expanding and in which more and new phenomena are becoming amenable to observation and analysis and new and novel features of such phenomena are being revealed, is an unlimited storehouse of problem areas beckoning research workers to come up with great discoveries and even inventions.

2.2 Choice of Problem Area

Sometimes a research worker may have thought over some problem of interest to him (her) and possibly of concern to some others which the researcher feels can be pursued by him (her) given the domain knowledge already possessed and/or likely/required to be augmented. It is also possible that the problem choice was motivated by a critical survey of the existing literature and/or the generally felt importance of the problem and/or the perceived appreciation of even a partial solution to the problem. There could be other factors, not excluding – somewhat unfortunately – the perceived ease of completing the research in a relatively short time.

More generally, however, a potential researcher first identifies a broad area within his/her subject of specialization that appeals to him/her as interesting or challenging, at the same time doable, may be with some guidance from a supervisor or adviser or senior. One then scans available literature in that problem area to gradually limit the area of possible research. Fortunately, this step has been greatly facilitated these days through easy access to databases and books as well as articles that have a bearing on the problem area chosen initially. In fact, one can even access documents which spell out open problems in different areas. One can conveniently come to know about researches being conducted by concerned research organizations to identify some research problems not yet taken up by someone else.

Problem areas emerge from discussions among peers, attentive perusal of editorials in newspapers, experiences gained by visits to places not seen earlier and even during the journey, thoughtful analysis of some statements by a leading scientist or economist or social worker, one's own urge to solve a long-experienced problem at home or in the workplace or in the society, and a whole host of similar other sources. The following is a small sample of such problem areas randomly chosen from different spheres just to illustrate how a potential research worker can embark on problem-solving as a research activity in the sphere of his/her own interest or expertise.

Problem Area 1

Different states/provinces within a country claim their 'investment potentials' to be quite high and it becomes difficult for a willing investor to choose a destination, based on such claims.

There is no consensus evidence-based index of investment potential currently.

NCAER got involved a couple of years back and took up the problem.

- To work out a State Investment Potential Index which would be based on evidence pertaining to several dimensions of this potential and would be acceptable to users.
- The objective was to rank the states by scores on the Index, possibly supported by an index for each important dimension.

Evidence would include

- Data on infrastructure including indicators of quantity and quality of land, of transport facilities, of banking and other financial services, of skilled manpower. etc.
- Information relating to existence and implementation of different relevant regulations such as labour laws.
- Perceptions of investors about political stability and governance, prevailing investment climate and the returns likely from investment in future, and so on.

Evidences were to cover a period of several consecutive years. Since the potential tends to increase overt time, a long base period is not justified.

Evidences were to be checked for validity and mutual consistency.

Information items and perceptions as also their impact on investment potential were to be duly quantified to allow combination.

Appropriate steps were to be taken to make the (measures) of investment potential unit-free and easy to interpret.

The Index finally derived was to be validated against real situations on investment in future periods of time as well as against similar though not identical measures like the Global Competitiveness Index or indices computed by rating and consultancy agencies.

Problem Area 2

We encounter a complex or an ill-posed objective function in an optimization problem. It is also possible that some of the variables and relations involved in the problem are ambiguous or inter-valued or fuzzy. Existing algorithms do not help in reaching the 'optimal' solution. At the same time, more and more real-life problems exhibit such features In fact, many discrete optimization problems in sequencing and scheduling and also in network analysis belong to this category.

The objective in such a situation is to develop an algorithm or a meta-heuristic to solve such a problem, even in terms of an approximately optimal or a weakly optimal solution. Genetic algorithms, simulated annealing, taboo search methods, ant-colony optimization algorithms and several other approaches have been offered recently to work out a 'satis-ficing' solution, though proving convergence of the solution continues to haunt research workers.

Problem Area 3

We notice a gradual shift in the foreign policy of a country from its initial stand on non-alignment, as indirectly revealed in statements made by the leaders in public and recorded in media or in written documents bearing on the country's relations with its neighbours.

The objective is to detect the direction and extent of shift during the last decade, say. In this case, we do not have the opportunity to interact with policy makers and leaders to get their views or statements. And whatever the sources of data may be, they will reveal a lot of disagreement and we will have to extract only those aspects which will be found to contain a reasonable extent of concordance. In such researches, Content Analysis has emerged as the right type of analysis of the categorical data that we can collect.

Problem Area 4

In many large-scale sample surveys, a multi-stage stratified random sampling design is adopted. The researcher has to work out a suitable allocation of the total sample size n (decided on a consideration of RSE of the parameter estimate and of resources available) among the strata with sizes N_i to get the sub-sample sizes n_i such that $N_i \geq n_i \geq 1$ and Σ $n_i = n$. The optimal allocation formula does not yield sub-sample sizes that are integers and that satisfy the other constraints always and there exists a problem of deriving some formula that can ensure that all the constraints are satisfied.

Problem Area 5

In dealing with highly complex and, at the same time, highly reliable systems, we cannot directly estimate the probability of system failure in the absence of even a single failure observed yet. We have recourse to fault tree analysis and represent the system failure event in terms of an event tree diagram based on Boolean logic and build up a somewhat involved relation connecting the probability of system failure with failures of lower-level events for which we can get some observations to estimate their probabilities. To develop such a relation we require to identify the smallest number of minimal cut sets and for this purpose we need to develop a suitable and efficient algorithm.

Problem Area 6

Environment and Ecology coupled with Development have become areas of hot pursuit by researchers from various backgrounds, sometimes working by themselves alone and in other cases coming to form interdisciplinary research teams. When one speaks of bio-diversity at different levels like eco-system diversity, species diversity and genetic diversity, one may feel tempted to develop a workable identification of an ecosystem or a biome, based on data relating to physiography or terrain features, rainfall, humidity and tempera-ture profiles, existing forms of life and their habitats as well as interactions and the like.

One obviously comes across multivariate data and the research problem is to define an eco-system based on available data and, using this definition, to identify different ecosystems within a given geographical region. An ecosystem must be internally homogeneous and distinguishable from any other (adjacent or distant) ecosystem. Another problem could be to develop a measure of bio-diversity in plants or animals in a given ecosystem that can detect any genuine loss in diversity.

Problem Area 7

An emerging area for research is woven around the topic 'intellectual capital'. In the context of production functions where 'labour' and 'capital' are the two input factors considered to explain variations in value of output, it is found that this value depends on the efficiency of deployment of the input factors and this efficiency, in turn, depends on the intellectual cap-ability of people within the organization. According to some researchers, this capability is not just the level of education or of skills possessed by the people or their analytical ability to understand systems and processes and to manipulate the variables involved based on this analytical ability but also their 'emotional intelligence'. In fact, the emotional intelli-gence of decision-makers and leaders, in particular, plays a major role in determining the value of output, given input factor levels. The issue that comes up is how we quantify 'intellectual' as a third input factor in the production function. Several alternatives have been tried out and more remain to be explored. Maybe no one single quantification will work out satisfactorily in all situations. At least, the role of intellectual capital will vary from people-dominated production situations to machine-dominated situations. A big problem area is to develop suitable formulations of intellectual capital for each distinct production situation type.

Problem Area 8

Coming to Health Research and specifically to quantification and use of 'burden of disease' in general or for a group of diseases which account for the majority of premature deaths or disabilities, the World Health Organization (WHO) has evolved standard procedures for evaluating disability-adjusted life years (DALY) and also related measures like years of life lived with the disease or disease-adjusted life expectancy etc. But a big issue remains to find out the weights to be associated with different disability categories. In most countries, data on disabilities associated with home-treated disease incidences are quite inadequate and lacking in credibility. And an altogether different but important study could evolve around the economic impact of disabilities of different sorts causing loss of man-days in production.

Problem Area 9

In psychological studies, mental tests covering a wide spectrum are used to describe, ana-lyse, predict and control different behaviours and actions. Some of these tests attempt to measure some inherent traits, while others try to bring out certain abilities. Many of the tests or inventories of test items are developed separately for different age-groups or levels of education or specific tasks/responsibilities. Inherent traits as revealed through observable behaviour or action and abilities to perform certain tasks or answer certain questions regarding numerical ability or abstract reasoning or language skill and similar other aptitudes are grossly affected by access to various devices and knowledge and skill improvement tools in recent times. Thus the adequacy or appropriateness of some test developed several decades back and, in some cases, partly modified in current revisions throws up a serious question. In the opinion of the author, a vast area of research is open for comprehensively assessing the psychological tests and their scaling procedures as well as interpretations in modern days characterized by digital devices and information flows along with new avenues for inter-personal interactions and the present-day social and economic environments in a globalized world that prompt people to think, speak, behave and act in particular ways much different from what prevailed during the days when the different tests were developed and subsequently modified.

Problem Area 10

Environmentalists regret the loss of bio-diversity at eco-system level, species (including sub-species) level and genetic level that have been taking place over eons of time. In fact, we are told that a large number of paddy plant varieties grown in olden days in parts of south-east Asia have become extinct nowadays. Life scientists who are more pragmatic and attach due importance to economic considerations in farming argue that farmers have consciously discontinued cultivation of these varieties despite the fact the corresponding rice varieties fetched higher prices than those which survive today. The reason behind this conscious decision was the fact that resources required for cultivation of these extinct plant varieties per unit of output were pretty high. The yield rate was low and more than offset the advantage in prices. To the farming community, other consequences of loss in bio-diversity do not weigh as much as costs of cultivation and value addition per unit area. A big research problem is to balance these two contrary considerations and, taking due care of the impact of bio-diversity loss on the environment, to identify which varieties are to be withdrawn and which to continue. Something like a non-zero-sum game problem, a satisfactory solution has to be found.

Problem Area 11

In different facets of industrial management dealing with total preventive maintenance or quality assurance or facility layout planning or customer complaint resolution etc. once a new practice (usually involving a new methodology) is introduced, its effective-ness is evaluated in monetary terms by considering the visible and direct gains which can be attributed to the new practice, putting under the carpet positive contributions from different ongoing practices in areas that impact on the gains from the new practice. This evaluation has an upward bias in figures on gains. It should also be examined whether the gains being attributed to the new practice under consideration would have other-wise been evident as a consequence of usually happening improvements in the ongoing practices only. Thus, going by the falsifiability criterion of Karl Popper, the current evalu-ation method is unscientific. A vexing and difficult-to-answer question will be how best to evaluate gains from the new practice, given a situation where ongoing practices cannot be kept the same and not allowed to change for the better, simply for the sake of evaluating gains from the new practice.

Problem Area 12

Of late, we talk about many indices that reflect the prevailing state of development (taken in a broad sense) in a country or a region and tend to compare countries and regions using some such index. Some of the oft-used indices have been developed by some international agencies and these naturally attract a lot of attention. However, most – if not all – such indices involve some amount of subjectivity and suffer from one limitation or another. Thus, these indices provide a big scope for improvement. Let us consider the widely used Human Development Index and the Human Poverty Index developed and released by the United Nations Development Programme (UNDP) team. The Human Development Index (HDI) is an unweighted or equally weighted arithmetic mean of four sub-indices, the viz Life Expectancy Index, the Educational Attainment Index (two parameters giving two sub-indices) and the Income (level of living) Index. While a lot of thought has gone behind the development of each sub-index taking due care of simplicity, data availability, convenience of interpretation, comparability across countries with different currencies, governance practices, cultures and the like, there are many grey areas which can be examined by research workers.

The HDI involves Gross Domestic Product (GDP) per capita expressed in Purchasing Power Parity (PPP) dollars or international dollars. The PPP dollars is conceptually defensible but computationally complicated and not so credible. This figure is adjusted to take into account the fact that an increase in GDP per capita beyond a certain threshold does not contribute to a proportionate increase in level of living as reflected in quantity and quality of goods and services consumed. Currently the natural logarithm of GDP per capita is considered, though earlier a more complicated formula due to Atkinson was used. It is true that we have to consider some monotonic function of income that has a slower rate of increase compared to income. But, which one should be used?

Coming to the first sub-index, expectation of life at birth is a summary figure that tells little about the distribution of age at death or length of life. And with the current emphasis on burden of disease, we may think of the Disease-Adjusted Life Expectancy (DALE). However, data on disability due to diseases are not available in many countries. The Human Poverty Index goes one step further to consider the probability of death before age 40. The choice of this age remains a question.

For educational attainment, adult literacy rate and Combined Gross Enrolment Ratio (CGER) should not suffice. The CGER is a great mix-up of disparate entities. Can we not consider just the percentage of people successfully completing 10 or 12 years of schooling or some such measure?

A common deficiency of all the sub-indices is that they assume a linear progression of the underlying parameter from the state of least development to the most developed state. It is known that as we progress towards the most developed state, more and more efforts are needed to achieve the same closeness to the ideal or most developed state.

We sometimes come across research being currently done or completed to effect improvements in some measures or methods of analysis for a vexing phenomenon that has been pursued over decades. One such phenomenon or problem area is related to inequality and poverty and more precisely to income inequality and income poverty. Concepts and measures of inequality in income distribution in a country or a geographical region were introduced nearly a century back, in the seminal works of Lorenz (1905), Gini (1914), Bonferroni (1930) and others. And the number of published contributions by research workers across countries has been well over one hundred. In the last few years the Zenga Index (1984, 2007) has been extensively studied by several research workers, working out its non-parametric estimate in terms of U-statistics and the asymptotic distribution of this estimate. Linked to this index is the question of defining the 'poverty line', which has been developed in various alternative ways including considerations of the distribution of carbohydrate consumption and the joint distribution of protein and carbohydrate consumption. Once a 'poverty line' is agreed upon as an income figure to enable a minimum standard of living, the issue of developing an index of poverty comes up and we have a whole host of numbers which satisfy some desiderata for a poverty measure to some extent or even fully. But then the development of neither a poverty line nor of a widely accepted and applicable index has been a closed chapter and research continues vigorously, patronized by potential users among economists, statisticians, social scientists, bureaucrats and politicians.

Problem Area 13

In several developing countries, the unemployment rate taken along with rates of under-employment and of labour underutilization reveal a somewhat gloomy picture about development in the broad sense. And one possible reason for this which should attract the attention of social science researchers in such and even other countries is the fact that gov-ernment policies in these countries restrict a certain proportion of jobs in each of several categories to candidates from specified social or economic or religious or ethnic groups. Such reservation policies might have been adopted by the corresponding governments to improve the employment situations and accordingly levels of living among some back-ward groups. However, one consequence of such reservation policies is the more than expected rate of unemployment in segments of the labour force which enjoy no reser-vation. In fact, for some jobs requiring minimum prescribed educational/professional backgrounds there could be sufficient job-seekers in the 'general' segments of the labour force who fit into these employment opportunities, but many of these jobs remain vacant due to their being declared as 'reserved' and there being a dearth of adequately qualified persons in the reserved categories.

A more challenging task could be to compare the 'cost' of vacant positions in the job market with the attendant unemployment in the 'general' segment of the labour force and spending resources to equip job-seekers from the 'reserved' segments of the labour force with the requisite education and skills, followed by withdrawal of job reservations. One can even study the social impact of reservations in jobs creating a sense of inequality in opportunities for employment and promotion etc. among different sections of the society.

An associated point to be examined is the possible mis-use of the reservation policy, in terms of children of persons from the 'reserved' segments who have secured good oppor-tunities to provide their children with good standards of education and training to con-tinue to enjoy the same privileges as their parents. And the benefits of reservation usually tend to help some individuals from the backward communities and not those communities as such.

One cannot comment on the desirability or otherwise of reservation policies in the employment market and of other feasible alternatives to provide more opportunities for self-development to members of the backward communities without making adequate studies on the above and related issues. It seems there are political or communal overtones in such studies or in proposals for such studies. However, research should steer clear of these overtones and provide inputs for better and more effective state policies for all-round development of the entire country.

Problem Area 14

In recent times, use of available prior information is being increasingly incorporated in the premises for any statistical induction relating to some aspect of an underlying phenomenon.

In fact, the Bayes methodology for making such inferences helps us in making more robust inferences in the face of model uncertainty including uncertainty about model parameters. Usually, prior information has been accommodated by way of a prior distribution assumed for model parameter(s) to compensate for lack of adequate knowledge about the model and the parameter(s) therein. And the choice of a suitable prior becomes an important issue. The prior distribution is used to derive the posterior distribution of the parameter(s) based on the likelihood of the sample observations assuming a given probability model coupled with the prior distribution for the parameter(s) considered.

There exist situations where prior information is limited only to change the support of the probability model assumed to represent the data: for example, in a series of experiments to improve the existing design of a process or a product, the prior information about the distribution of a feature that is assured to increase (decrease) in each successive experiment as a result of the design improvement exercise. For example, in such an experiment to improve product reliability, we may be using a simple one-parameter exponential distribution for failure time over the support $(0, 236)$ initially. After the first trial the support may be shifted from zero to $k \lambda$, where λ is the mean time-to-failure found during the first trial (maybe estimated population mean time-to-failure) and k is a safety factor so taken as to ensure that no time-to-failure observed during the second trial goes below $k \lambda$. We may not have enough evidence to assume a prior distribution of the parameter λ.

A problem area for workers in many disciplines arises from the fact that several response variables have been observed and the need is to combine them suitably into one or two linear combinations so that most of the information contained in the response set is extracted, the relative importance of each response variable in the context of the objectives of the study is preserved, the combinations can admit of reasonable explanations, and application of relevant statistical methods and tools remains valid.

The task of projecting the non-Euclidean surface of the Earth on a Euclidean plane (of the map) so that relative distances between locations on the Earth's surface are as close as possible to the Euclidean distances has been a research problem with cartographers since the days of Mercator. There have been several developments beyond Mercator's projection, but the problem of approximating relative distances to better-than-before exercises continues. In fact, this is virtually a mathematical problem in mathematics related to projection of one space onto another with different objective functions. The problem can be looked at as one in multi-dimensional metric scaling, a tool that arose in connection with perceptions in psychology and was solved in terms of an algorithm provided by a psychometrician.

A wide green pasture for research workers lies in the field covered by the somewhat odd phrase Industry 4.0, which is expected to create some irreversible impact on our lives, changing the nature of the work we do and the nature and location of work-places in future. This corresponds to the digitally enabled fourth industrial revolution to come, propelled by artificial intelligence, the Internet of things, block chain, big data, machine learning and other related advances in science and technology. We are flooded with data arising from a host of sources at an incredibly fast rate corresponding to the huge volume of different types of transactions carried out by us, not all structured and ready for analysis, not all of equal value and not all equal in veracity. We have ourselves developed huge capabilities for processing large volumes of data in fractions of a minute. We try to connect objects and operations through sensors. With all this, we often fail to predict the future. Some ascribe such failures to inaccuracy or inadequacy of data, while others argue that the models used for predictions were wrong in the given context. Some even realize the limitations of algorithms and machine hardware as possible reasons. Apart from the challenges posed by growth in artificial intelligence, predictions based on big data for such events as earthquakes or tsunamis or eruptions of volcanoes could be useful areas of research to benefit mankind as a whole.

Another area of research whose findings are becoming more and more relevant concerns human rights and their relation to development (essentially material or economic). In fact, social scientists argue – and rightly so – that as we achieve higher states of material development, we tend to violate several human rights. Some research of an unconventional type has attempted to establish some relation between violation of the most important fundamental human right, viz. the right to life, and state-sponsored terrorism or, at least, failure of the state to contain murder or failure to protect workers in risky development projects from fatal accidents. In contrast, technologically sound development may also imply fewer casualties resulting from violence or from accidents.

Research revolves round research problems that await solutions. And many research problems arise in the interface between different disciplines. Some problems were looked at by the present author – though no research should be claimed – while interacting with scientists and engineers. Mixture experiments with constraints on the mixing proportions, recognizing randomness in the estimated response surface in any attempt to locate the 'true' optimum, establishing seasonality in monthly data over several years, estimating interaction and even main effects in simple factorial experiments to rule out some treatment combinations as not scientifically tenable, using results in game theory to find parameters in a single sample acceptance-rectification inspection plan, trying out mode-minimizing solutions to some stochastic models in operational research and a host of similar ventures reveal the need for inter-disciplinary research.

Collusion, corruption and unethical practices in public procurement of materials, equipment and services (including those of consulting firms and inspection agencies) and in execution of public projects with their consequent impacts on quality of performance (including safety aspects) and time and cost over-runs are not uncommon problems of concern to public administration. Apparently elusive for analysis, this problem has been studied by using grounded theory (Locke, 2001) to enumerate and code different events happening during contracting, procuring and execution followed by application of fault tree analysis to link events at different levels and failure mode, effect and criticality analysis to prioritize the different modes of failure (top event) based on estimation of the small probabilities involved. Interesting application of dynamic game theory has also been attempted to find optimal strategies to minimize the risk of the public administrator as one of the players with the vendor as the other in situations where an inspection agency is hired by the administrator (who may collude with the vendor). This is a sensitive problem area for research in public administration with challenges in data collection and imaginative application of methods and techniques used in industrial engineering and operational research.

By far the most important meta-level research problems which call for credible, adequate and timely data on different dimensions of human activity followed by their appropriate analysis and which are likely to impact our lives and our planet earth are the ones associated with development as a multi-dimensional process along with the outcomes thereof. Much after some of the attempts to quantify human development came to be accepted and the relevant findings in many countries were found to be quite disappointing, the United Nations announced a set of Millennium Development Goals (MDGs) along with corresponding measurable targets to be achieved by member countries by the end of 2015. Later, a set of seven Sustainable Development Goals was announced and adopted. The 17 Sustainable Development Goals declared in the UN Summit 2015 to be achieved by the end of 2030 aim to end poverty, promote prosperity and people's well-being and protect the environment.

The 17 goals set out for the next 15 years are a blueprint for a better future. These are

- Zero hunger
- Good health and well-being
- Quality education
- Gender equality
- Clean water and sanitation
- No poverty
- Affordable and clean energy
- Decent work and economic growth
- Industry, innovation and infrastructure
- Reduced inequalities
- Sustainable cities and communities
- Responsible consumption and production
- Climate action
- Life below water
- Life on land
- Peace, justice and strong institutions
- Partnerships for the goals

Extreme poverty was taken in the context of MDG's as implying an income below USD 1.25 per person per day.

It is worth mentioning here that the extent to which these goals have been achieved in any country since they were pronounced is determined in terms of as many as 232 indicators and in many countries some of these indicators have to be derived from data on substitutes. While a lot of conceptual research backed with limited information about actual situations in most of the developing and less developed countries has gone into the formulation and promulgation of these goals, no commensurate research has been taken up in any country. In fact, most countries get figures for the parameters in each goal from their respective statistical offices and compute the extent of progress towards achievement of any goal. Research on policy and planning, projects and their deliverables in the fields of agriculture and food processing, manufacturing and allied activities and, more importantly, in the sector of services like education, employment, health care, environment and the like – aligned towards the development goals – are sadly lacking. Some scanty data analysis and an attractive report by some consulting agencies do not constitute the desired research.

Just to mention one problem area for research to ensure adequate food production, assumed to be followed by effective distribution, concerns land use planning. Let us for the time being restrict our attention to agricultural land use planning only with no foreseeable stakeholder other than the crop production system. If agricultural land use planning has to address the issue of food and nutrition security – as is naturally expected – we have to work out the minimum area of land to be kept under cultivation and food crops to feed the growing population of the country in each region/part. This minimum must take into account the best possible utilization of land to be available for food production with the best achievable farm management practices. The concern for food security is primarily a concern for yield, with no direct and necessary linkage with economics of production and hence with livelihood/income levels of those engaged in food production. This security being a major national interest, a national policy of land use indicating broadly the allocation of the total available land for different identified and competing use like food production, production of commercial crops, infrastructure development, urbanization with its associated ancillary requirements, development of industries and commercial establishments, etc. is a bad necessity in many countries.

To develop land use plans at district levels for land currently under agriculture and likely to remain so in the foreseeable future, the following distinct activities are implied:

1. identification of alternative uses of the land parcel (a relatively big chunk with more or less homogeneous soil characteristics) under consideration in terms of

   (a) cropping systems like multi-tier cropping, multiple cropping, relay cropping etc.

   (b) within each feasible cropping system crops and crop sequences;

2. recognition of constraints like availability of water, soil characteristics, investment capability of the farmers and subsidies available to them, changing climatic conditions in the area, farmers' access to extension services, marketing of crops, and the like;

3. exploring possibilities of constraint removal or reduction through technological innovation, interactions with agricultural banks and other agencies like the agencies for rural electricity distribution etc.

2.3 Formulation of Research Problems

Once the research worker has identified the problem area or even the research problem, what is needed to proceed further is a concrete – fully or partly – formulation of the problem in the language of the context. One should state what exactly one is aiming at by way of a complete or partial solution to the problem, following some established approach or suggesting and adopting a new approach towards solving the problem, verifying some existing results in terms of relevant evidences accumulated since the existing results were reported, coming up with a new interpretation of some earlier result or finding looking at the context from a different angle, simplifying the existing derivation of a result, relaxing some existing restrictive assumptions in the existing model in establishing some finding, and so on.

Formulation of the research problem reveals a wide spectrum. In sponsored research, the problem has already been identified by the sponsoring people and has also been formulated in some details which may require further amplification, concretization and focus on what is meant by the 'intended solution'. The intended solution may be too ambitious in terms of its cost, scalability and utility features and may require some pilot experiments followed by a re-look at the initial characterization of the intended solution. In some rare situations, the sponsoring agency may be quite modest and the pilot results may even suggest some improvement over the initially expected result or solution. Sometimes research problems are identified out of a box of problems suggested by some national agency to meet some planned targets for improving socio-economic conditions in the country or to preserve or enhance the quality of the environment. A comprehensive perusal of relevant literature in some chosen problem area and an assimilation of related experiences may help finding out the deficiencies in knowledge to describe, analyse, control and optimize some phenomenon so that the research worker/group can come up with a clear formulation of the research problem. There exist some situations where, of course, research workers have full freedom in identifying their research problems based on various inputs including their own interests in knowledge acquisition and in service to humanity.

Problem formulation must take due account of resources – material as well as human – currently available or likely to be augmented. Of course, resource sharing among institutions or among research groups should not be a problem necessarily.

Validation of research results is a crucial step and is sometimes compromised in terms of a limited dissemination of the results to some selected stakeholders with a view to getting their honest feedback.

2.4 Role of Counter-Examples and Paradoxes

The history of Science as a body of knowledge arising out thinking about strange or uncommon or even complex entities and related problems that seemingly evade any solutions has been characterized by sophisms, paradoxes and counter-examples to established results of previous workers. Even conjectures made by some imaginative individuals have continued to await complete and convincing proofs. In fact, it has been argued that most choices made (and thus decisions taken) and actions implemented are marked by volatility, uncertainty, complexity and ambiguity. Thus, new and novel approaches are required to render our decisions and actions robust against any of these disturbing elements. And thought-provoking researches have been motivated by these elements. And once a research worker puts forth some new result, even a single pathological example that disproves the same initiates further research. Some paradoxes simply point out the fact that solutions are fragile, their validity may be limited and it will be unsafe to claim anything 'big' in favour of any such solution.

A sophism is an argument which, though apparently correct in form, actually contains an error which makes the final deduction absurd. A well-known sophism is that "what you did not lose, you possess. You have not lost horns, hence you possess them". A paradox is a statement that seemingly contradicts common sense, yet in fact is true. The study of many paradoxes has played an extraordinary role in the development of contemporary physics. Consider one example. Linear dimensions of bodies change with temperature according to the law $l_t = l_0 (1 + \alpha t)$ where l_t denotes length at temperature t. Suppose temperature drops to $t = -1 / \alpha$, then the length becomes 0. Is it possible.to decrease temperature below this level to make length negative? Absurd. This led to a deeper study and it was found that even if the coefficient of thermal expansion is high, the value of $-1 / \alpha$ is much less than $-273°C$. For example, lead has $\alpha = 3 \times 10^{-50} C^{-1}$. In this case $-1 / \alpha$ is a temperature which can never be reached. We now realize the dependence of the coefficient of thermal expansion on temperature, too.

It is well accepted that the coefficient $\beta_2 = \mu_4 / \mu_2^2$ measures the 'kurtosis' or 'peakedness' of a probability distribution curve compared to the normal distribution with the same standard deviation. A value of this coefficient greater than 3 is taken to imply that the curve under consideration is more peaked than the corresponding normal distribution and similar is the interpretation of a value less than 3. However, Kaplansky (1939) came up with a counter-example in terms of the probability density function f (x) = with $\beta_2 < 3$ but with the highest density exceeding $1 / (2 \Pi)^{1/2}$. While it is true that the value of β_2 correctly indicates the flatness or peakedness of most distributions, this single counter-example should motivate a search for a foolproof measure of this property of a distribution.

The St. Petersburg paradox which influenced later thoughts of economists was posed as a question by Nicholas Bernoulli (1713), as stated by Hald (1984), which can be stated as follows: consider a game between two players, Peter and Paul, in which a fair coin is tossed repeatedly until the first head appeared. If this was at the jth toss, Paul would receive an amount 2^{j-1} from Peter, j = 1, 2, ... How much should Paul pay initially to Peter as an entry fee to make the game fair or equitable? The principle that a fair entry fee for a game should equal the expected gain of the player had been accepted earlier. In this case the expected gain to Peter works out as $\Sigma^{\infty}_{j=1} 2^{j-1} (\frac{1}{2})^j = \Sigma 1 / 2$, which is not finite. And we cannot argue that Paul should pay an abnormally high entry fee for the game. The problem was solved by Daniel Bernoulli (1738) using the basic argument that the moral worth or utility of an absolute increment in the wealth of a person depends inversely on the amount of wealth already in possession. Denoting utility by y corresponding to a nominal wealth x, he stated that dy = c dx / x. This meant that a person's utility increases by $\delta y = c \log [(\alpha + \delta x) / \alpha]$ where δx indicates the increment in nominal wealth possessed. On the basis of this concept, the expected gain in utility for Paul would be

$$E (\alpha) = c \Sigma 1 / 2^j \log [(\alpha + 2^{j-1}) / \alpha]$$

Implying a gain in monetary possession e $(\alpha) = \pi (\alpha + 2^{j-1})^{1/2} - \alpha$ with values e (0) = 2, e (10) ~ 3 e (100) = 4 approximately and so on. Bernoulli's argument has since been accepted and adopted by economists.

The expected utility criterion to compare two alternatives can lead to a situation where the decision-maker does not make a choice or settles for a 'mediocre' alternative. Let us consider a simplified version of the Ellsberg Paradox. There are two urns filled with two balls – black and red – each: the first is known to have one black and one red ball. In the second urn the unknown proportion of black balls has a uniform distribution. The bet is to draw one ball either from urn 1 or from urn 2 and if it turns out to be black, the drawer wins USD 100. From which urn does one draw a ball? It can be easily verified that the probability of getting a black ball from urn 2 is 1/2, the same as that for urn 1. However, it is generally found that most people draw a ball from urn 1, possibly because it has a given probability 1/2 for winning the bet, as against an unknown probability for urn 2. Most decision-makers are risk-averse and want to avoid uncertainty. In fact, one could get a white ball from urn 1 and a black from urn 2 and win the bet. Thus decision-makers settle for a mediocre solution. This paradox has given rise to several versions of the theory of uncertainty offered by different investigators. However, a choice between probability theory and uncertainty theory does involve uncertainty, which never leaves us.

2.5 Illustrations of Problems

The following is an illustrative set of problems mostly in the area of statistics, some of which have already been solved and the others still await solutions.

Problem 1

Fact: Parameters in a probability distribution are estimated using moments of integer orders.

Question: Do we lose or gain in efficiency by considering fractional moments?

Answer: We can gain by removing the integer constraints and determining the orders (may be fractional) of the moments to maximize (relative) efficiency of the moment estimators. It has been tried out in the case of a two-parameter Weibull distribution.

Problem 2

Fact: In designing an experiment with n factors each at two levels (usually a higher and a lower one), we assume all possible 2^n treatment combinations as feasible and proceed to estimate factorial effects –main and interaction.

Question: What happens if some level of one factor is not compatible with some level of another factor?

Answer: Estimation of factorial effects is difficult.

Problem 3

Fact: The (unknown) *true value* of a measurand is estimated as the mean of some repeat *(observed)* measurements.

Question: How much is the *uncertainty* about the *true value* and how do we estimate the *(standard or expanded)* uncertainty?

Answer: The standard uncertainty is defined as the standard deviation of the (unknown) *true values with which the observed value or the observed mean value may be associated and is estimated from a model as an asymptotic result.*

Problem 4

Fact: To test H_0 ($\sigma_1 = \sigma_2$) in a bivariate normal distribution for (X,Y) against both-sided alternatives, we use the result that $U = X + Y$ and $V = X - Y$ are uncorrelated and reject or fail to reject H_0 using the t-test for zero correlation between U and V.

Question: How do we take into account other possible consequences of H_0?

Answer: Has not yet been examined.

Problem 5

Fact: Sample size in a survey to estimate a proportion π is often found by taking the relative standard error as 0.05 or 0.01 or 0.10 as

$$1 / \pi \sqrt{[\pi (1 - \pi) / n]} = 0.05, \text{ say} \tag{1}$$

Question: This formula applies to simple random sampling while in many surveys the design could be stratified random or systematic or PPS sampling. Should we use this formula or a larger or smaller n?

Answer: We should multiply n from (1) by the *design effect*, viz. $\sqrt{[r(n-1)]}$ where r is the intra-class correlation and n is the average size of a cluster. We should remember, however, that the sample size to be used in a survey does not depend only on a consideration of the relative standard error of the estimate for the parameter of interest. It does depend on resource availability and other aspects of field work involved.

Problem 6

Fact: In a single-parameter survey with stratified random sampling, we can divide the total sample size across strata using proportional or Neyman or optimal allocation.

Question: How do we solve the allocation problem in a multi-parameter survey?

Answer: We can use the proportional allocation formula. But for the other formulae, we do not have a unique extension. Research has been going on.

Problem 7

Fact: To find the optimum treatment combination in a factorial experiment, we optimize the (estimated) response given by a (polynomial) regression equation using some search algorithm.

Question: How do we take care of sampling variations in the *estimated regression coefficients* in the optimization exercise?

Answer: We can possibly use stochastic optimization algorithms. Somewhat difficult.

Problem 8

Fact: In a Bayesian estimation exercise, we often choose natural conjugate priors for the parameter(s), so that the posterior has the same form as the prior.

Question: Does this prior truly reflect prior information about the parameter(s)?

Answer: It may be better to use maximum (subject to some rational constraints) entropy priors or the partially informative impersonal priors introduced by Jaynes (1988). Of course, when the variable X under study follows a distribution in the exponential family, the natural conjugate priors have the same form as the impersonal subjective prior maximizing entropy.

Problem 9

Fact: In estimating coefficients β's in a linear regression of Y on, say, $X_1, \ldots X_k$ we often tend to ignore some pertinent constraints like $\Sigma\, r_i\, \beta_i = c$ (constant).

Question: How to include such a constraint or an inequality constraint like $\beta_r < (>) \beta_s$ which may hold exactly or with some probability?

Answer: We can use Stein-rule estimators under exact restrictions and some mixed estimators in the case of stochastic restrictions.

Problem 10

Fact: In analysing bivariate data we usually consider distributions with marginals having the same form.

Question: How to proceed when the marginal distributions belong to different families?

Answer: We can generate bivariate distributions with different forms of marginals which can represent features of the two random variables under study using the Farlie-Gumbel-Morgenstern approach or some other approach.

Problem 11

Fact: It is known that Karmarkar's algorithm and, for that matter, ellipsoidal and interior point algorithms are weakly polynomial time algorithms for solving a large-scale linear programming problem.

Question: Does there exist an algorithm that is strongly polynomial time in the number of constraints and the number of variables?

Answer: This remains an unsolved problem to date.

Problem 12

Fact: For a two-state reliability analysis, we have standard measures of structural importance of different components in a coherent system in terms of path sets and cut sets.

Question: If we think of three or more states of components and systems, how do we measure the structural importance of components? Even the notions of path sets and cut sets are not standardized.

Answer: This is still an open problem in reliability engineering.

Problem 13

Fact: While interpreting the score obtained in a psychological test, we sometimes consider the percentile scores as given by the expert who developed and standardized it. He or she considered a group of individuals to whom the test was administered and whose trait or ability under consideration could reasonably be related to percentile score ranges.

Question: Should we accept the same ranges for percentile scores while interpreting scores obtained by individuals in a different group?

Answer: Possibly not. However, the real answer may depend on the nature of the trait or ability being measured and the way the test was developed and standardized.

Problem 14

Fact: Paris-Erdogan and Forman equations have been proposed to predict the growth of a crack due to fatigue in a metallic structure subjected to varying stress cycles. These contain parameters which depend on frequency of loading, temperature to which the structure is exposed and the ratio between the maximum and the minimum stress in the stress cycle or the critical level for the stress intensity.

Question: Given that these parameters are really not fixed for a particular metal and/or type of structure but are likely to vary randomly with variations in several different factors affecting their values, can we not attempt some stochastic version of these equations?

Answer: Yes, randomized versions of these models have been attempted. The approach has been to allow the parameters to be random with some assumed probability distributions.

Problem 15

Fact: There are countries where only a few population censuses have been conducted to date. In one such country, the first ever census was conducted in 2002 and the next one was planned to be done in 2014 but was actually held in 2016. Based on the census count of 2002, the population of the country was being projected annually using the component method, and the gross domestic product was being computed year after year to provide the GDP per capita figure as an indicator of economic development. The ratio between GDP computed for 2016 divided by the projected population for 2016 was found to differ a lot from the ratio with the census population for 2016. GDP computation need not be questioned. It is known that census population totals are usually correct and thus the discrepancy can be attributed to population projection not providing the correct population total. And the problem therein could be in terms of assumptions made regarding the future trends of mortality and fertility and even of nuptiality.

Question: It will be desirable to have a time series on GDP per capita, starting with the year 2002. How do we reconcile the two figures for 2016 to get a smooth series?

Answer: For 2016 one possibility is to use some osculatory interpolation formula by computing figures beyond 2016. There are several graduation formulae used to smooth demographic data and some such formula could be used. A somewhat different approach would be to re-visit the projection exercise and make more realistic assumptions. One can explore other possibilities for eventually coming up with a smooth series.

Problem 16

Fact: Customers or users put up complaints about products or services acquired or procured by them, right at the time of deploying or using them or sometime during the useful life of the product or service. Not all these complaints are accepted by the producer or service provider and sometimes they just communicate to complainants accordingly without attending to the customer by examining the product or service. Only those complaints which are accepted as 'failures' by the producer or provider are attended to. The basis for such a policy is that sometimes customers or users put up complaints willingly (in anticipation of a replacement) or unwillingly (in the absence of adequate knowledge about how to use the product or service) At the same time, a famous exponent of management theory advocates "Treat every complaint as a failure", considering false complaints as consequences of inadequate customer education provided by the producer or provider.

Question: How does the producer or service provider decide the genuineness or otherwise of a complaint from a particular user or customer without paying a visit to the user or customer?

Answer: A blanket answer does not exist and one has to make use of available data of different sorts to find an appropriate answer for a given complaint by a given person.

Problem 17

Fact: Several measures of effectiveness of equipment maintenance have been offered by research workers, some suiting the needs of works management, some applicable to control, some required by top management and some fitting the context of total preventive maintenance (TPM). Some of these measures are based on times, some on costs, some on energy consumption per unit of consumption, etc. Each measure has its own merits and limitations. Some of the measures invite a certain amount of subjectivity e.g. in defining quality rate as the last component of over-all equipment effectiveness (OEE) in TPM, we talk of value-adding time as a fraction of total operating time. How exactly to quantify value-adding time remains a problem.

Question: With acceptable conversions of time and energy consumed into costs, can we develop some effectiveness measure that will help us in identifying deficiency in the equipment which can be removed or reduced?

Answer: One school of thought would suggest that a single measure should not serve all purposes and should not combine all different aspects of the equipment. Otherwise, there exists enough scope for research to come up with a measure that will reconcile demands of various stakeholders to the extent possible.

Problem 18

Fact: In problems of discrimination among groups (racial, religious, social etc. or plant or animal taxa) and classification, we bring in the concept of a distance between two groups based on a number of features or characteristics observed for each of the groups. The D^2 statistic proposed by Mahalanobis has been widely accepted as a measure of distance. Of course, there are divergence measures between probability distributions like the Kullback-Leibler or Bhattacharya distance.

Question: These distance or divergence measures generally consider features or characteristics which are continuous variables. It is natural that in some cases some of the observed features are continuous variables, while some others are discrete variables. How do we develop a measure of distance based on a mixture of continuous and discrete variables?

Answer: Bar-Hen and Daudin (1995) have generalized the Kullback-Leibler measure of divergence to the Location model for this purpose; assuming that the discrete variables follow a multinomial distribution and conditionally on the multinomial vector, the continuous variables follow a multivariate normal distribution and dispersion matrices for the two groups are identical. One may look at other possibilities.

Problem 19

Fact: Of late, emphasis has been put on emotional intelligence as an important factor in life success, implying ability to accept and discharge responsibilities at home, at the workplace and in society or wherever an individual is positioned within a group. Components of emotional intelligence like sensitivity, maturity, empathy, resilience and competence are well recognized. We speak of emotional quotient (EQ) as a measure of emotional intelligence. There exist several trait-based and ability-based inventories to assess emotional intelligence.

Question: EQ is really not a quotient. We can use an EI inventory, use the appropriate scaling procedure and assign an EI score to an individual. This will serve the role of the numerator in a quotient. But what can be a corresponding denominator? We recall that intelligence quotient (IQ) is defined as the ratio between the mental age and the chronological age. What can be an analogue of chronological age in the EI context?

Answer: Not much has been done, on the plea that a quotient is not that badly needed in any scientific study to relate EI with some aspect of performance and the term 'quotient' need not be emphasized. Remains an open problem.

Problem 20

Fact: With recent advances in information and communication technology coupled with emerging developments in artificial intelligence, robotics and the Internet of things, it is apprehended that the future workplace and the nature of work for many of us will undergo a revolutionary change or a complete disruption of the extant pattern. It is true that drastic changes usually take longer to settle down. In some contexts, this change-over time may be smaller, whereas in some other contexts the hang-over of the present will last longer, delaying the absorption of the new and causing undesired consequences in a globalized world with countries embracing the dramatic changes impacting others.

Question: Can we guess the time by which a radical change will completely overhaul the workplace and change the nature and method of work in a given context?

Answer: To the best of information in the public domain, an analytic exercise to predict this time has not yet been taken up by any group, possibly with the tacit understanding that such a historical change will take its own time.

Problem 21

Fact: People across the globe are worried about violation of human rights. Right to life is the fundamental right that implies no loss of life due to homicides. It is also true that many people get involved in accidents and some lose their lives. In some cases, such accidents are associated with the collapse of a structure like a bridge or a building or with passenger vehicles overturned in the process of negotiating wretched road conditions or a collision between two trains coming from opposite directions. In none of these odd and unfortunate situations did the ultimate sufferers show any lack of care or caution.

Question: Should such deaths be accounted for in the ambit of violation of the right to life?

Answer: In current practice, such deaths are excluded from the purview of rights violation. In the case of accidents caused by negligence on the part of the state in construction, operation and maintenance of conveniences and utilities for the public, certain compensations are sometimes announced by the state agencies or governments. This can be construed as an acceptance of violation of the fundamental right to life, though implicitly at best.

Problem 22

Fact: Life-cycle cost analysis has become an accepted practice in comparisons between alternative models of an equipment to be purchased and deployed. Most often, life-cycle cost is calculated by adding to the cost of acquisition (purchase cost), to be paid at the time of acquisition, costs associated with the use of the equipment (e.g. cost of energy consumed during use, cost of manpower required to run the equipment, etc.), costs of maintenance and costs of disposal less the salvage value or the resale price at the end of the estimated useful life period. Such an equipment is usually meant to be used over a reasonably long period of time and to fail to perform satisfactorily on certain occasions.

Question: Associated with a failure of the equipment during use will be some costs which may not be negligible in all cases. A sudden breakdown may cause loss of production time, damage to in-process material, injury to the operator and the like. How do we recognize such costs in the computation of life-cycle costs?

Answer: Most such equipment and their important components must have been tested for their failure patterns and should carry labels to indicate failure rates. With the help of these rates as declared, one can estimate costs of failure and incorporate those within the life-cycle cost.

Problem 23

Fact: Air pollution in many urban areas has increased significantly over the years and concentrations of several toxic particulates have exceeded corresponding permissible limits in many cities and towns. Various factors like automobile emissions, use of firecrackers and explosive materials, emission of dust from construction works and the like have contributed to this unhealthy and even hazardous situation. The incidence of respiratory diseases has risen along with increased incidence of other diseases caused and/or aggravated by air pollution. Various estimates of the number of casualties attributable to effects of air pollution are sometimes reported. A reputed medical journal quoted a very alarming figure for the death toll due to air pollution in India every year.

Question: Death certificates do not always report the antecedent cause(s) of death except the immediate cause. And deaths from pulmonary failure or associated COPD may be attributable to other diseases or deficiencies from which the deceased were suffering. How can some deaths be related directly and solely to adverse impacts of air pollution? Further, if a certain proportion of deaths in a certain study area has been unequivocally attributed to air pollution (in terms of concentrations of some toxic particulate matter), how desirable or even rational is it to use that proportion to estimate the number of deaths in a much larger area or population?

Answer: Differentiation between the primary and leading causes of death has to be incorporated and adequate attention should be paid to analyse all relevant covariates of the deceased to account for vulnerability to the immediate cause of death. Type studies on small scales are being attempted by research teams in many contexts and estimates for large populations should be noted with due caution.

Problem 24

Fact: Many research studies including some highly innovative ones have been taken up by a host of researchers in various countries to convert traditional wastes resulting from production and consumption of raw and semi-finished materials as well as finished goods at different stages of their life cycles into 'wealth' – something that provides inputs into production of goods and services and, in this way, adds to national income. Vegetable wastes have been gainfully utilized in preparing composts for use as manure and for energy generation. Coming to a 'finished' product, viz. food, we come across a whole lot of wastage – planned and intentional as well as unplanned and unintended. We have hotels and restaurants which serve more than what can be and is actually consumed. Of course, food wastage also takes place when the number of guests turning up on a ceremonial occasion falls much short of the expected number for whom food was arranged. There have been stray attempts to dole out the left-overs or unused portions to hungry mouths who cannot afford even a single meal every day. Occasionally, extra food already served but not fully consumed or not yet served is stored in a refrigerator to be used sometime later. One wonders what happens with partly unused food plates collected from airline or train passengers. Food wastage on the one hand and hunger on the other co-exist in many situations.

Question: Can something be done to collect unused or wasted food properly for conversion into useful inputs into some manufacturing industry or even in agriculture or in the cultivation of bacteria for possible use in the pharmaceutical industry?

Answer: Some stray attempts are possibly being made to add value to wasted food – maybe segregated as completely unused, partly used and preserving the other part as such and partly used in a way that makes segregation difficult – and used for identified purposes. However, an established practice on a wide scale has not been known. An inter-disciplinary team of food scientists, medical scientists, bio-medical engineers and material scientists and similar other workers should focus attention on this problem, as distinct from the problem of avoiding food wastage.

2.6 Concretizing Problem Formulation

Just a vexing question or a conjecture or a fuzzy solution to a difficult question may not qualify to be the problem for some research. In fact, how to concretely state the research problem(s) in a specific research study is by itself a research problem on a different scale, of course. Should we put some interrogative statement(s) to be examined as the research problem? Or, should it be a collection of affirmative propositions (which could be a single proposition in some cases) which have already found some evidential support? Or, do we start with a bunch of negative statements putting at naught claims or findings of some earlier research? Or, could the starting research problem contain some mix of these three types? Commonly accepted practices rather than standards or norms are sought. And such practices may reveal variations across fields of research.

A part of the concrete formulation of the problem in evidence-based or empirical research is to indicate the extent to which the research findings are to validly apply. There could be situations where evidences to be gathered and analysed in a research investigation are not expected to apply beyond the scope and nature of evidences designed to be gathered. On the other hand, evidences to be gathered may correspond to a wide aggregate showing a high degree of homogeneity among its different parts or constituents. Given this domain knowledge, the research problem may include testing the validity of the research findings for the larger aggregate from which the evidences to be gathered would constitute a sample.

The above issue is not much relevant in theoretical research. A result obtained with due diligence and appropriate logic by way of a deduction would be applicable all the time, while results on evidence-based induction would attract uncertainty about their generalizability. In theoretical researches, a concrete formulation of the research problem should include any restrictive assumptions under which propositions or questions would be examined. For example, in tackling a vexing problem in pure mathematics, the problem formulation should clearly specify the nature of space, the type of operation, and similar other considerations under which the result(s) intended by the researcher will be examined

In empirical research, the problem formulation has to necessarily avoid any confusion about the connotations of the problem so that the research worker and the beneficiaries of the research results can match the results with the connotations to judge whether the connotations have been fully met by the research work. And right in this context, a clear statement about the scope of research becomes a must. To illustrate, let us consider an important issue in pedagogy. Which of the following methods of teaching a subject is better? (a) Starting with an example or a case study or a simple problem to be answered, we introduce the topic in its general form for subsequent discussion. (b) Starting with the general statement of the material content in the topic, we come to consider some illustrations. It is well appreciated that the choice – a research problem in pedagogy – cannot be generic and would depend on specific situations characterized by (1) the nature of the subject and topic, (2) the level or grade of participants and (3) criteria to make a choice, such as ease to remember, convenience to grasp, opportunity to apply or use knowledge and/or skill acquired. Therefore, for a particular research on this matter, the research problem statement must provide these specifications. Of course, an ambitious research could examine results under different set-ups, study differences in results and provide relevant explanations.

For the very mundane requirement of adjudication of a research report (for an academic degree in an educational institution or certification of acceptance as a basis for action), a clear, unequivocal and complete formulation of the research problem has to appear in the beginning, immediately after the context, and whenever desired research findings should refer to the problem statement to some extent or other.

To emphasize 'concreteness' in formulating the research problem, we may keep in mind two aspects of 'uncertainty', viz. *vagueness and ambiguity*, which should be avoided. Vagueness arises from a fact like the domain of interest is not delineated in terms of clear or sharp boundaries (Klir and Folger, 1993). Ambiguity traces its origin to the inability to make unique choices in situations of one-to-many relations. Thus, to develop and prove characterization results for some probability distributions, one may preferably spell out the domain of interest as, say, the class of all uni-modal continuous distributions. In a socio-economic survey based on households, the domain of interest may, if so needed, be restricted to households for the civil population only. To avoid ambiguity in reckoning the employment status of an individual, one can clearly circumscribe the employment status as employment 'for pay or profit' if that is what is to be brought out in terms of findings of a household-based survey on employment-unemployment.

While all the above points draw our attention to the need for and importance of a concrete formulation of the research problem as the starting point and the guiding force in a research, it has to be noted that even when such a formulation has been developed and a research design worked out to proceed further, the possibility that evidences ultimately available and appropriate analysis carried out on such evidences do not do justice to the problem formulation. And a possibility of this sort is likely to be encountered more often in theoretical research than in empirical research. There could be several explanations for such a possibility. The problem formulated defies a convenient solution. The formulation took for granted some conditions or results which eventually turned out to be not applicable or not valid. The amount and quality of evidences called for could not be gathered with the resources available; for example, data on water quality were to be gathered for different parameters of water quality from a large number of water quality monitoring stations in several states through which a river under study flows. Coming to actual data records, it was found that the parameter pH is measured by using a pH meter in some stations and using pH paper in some others. And data from these two practices could not be pooled together.

To be noted is the crucial fact that every step in the research process including the initial step of formulating the research problem (which could involve several sub-problems) may justify a revisit, either during that step or later in terms of what could be done during that process on the expected outcome or working hypothesis. And, no wonder that in some researches such a revisit would need a re-formulation of the research problem in all its relevant details.

It is sometimes argued – and not without reason – that the justifiably expected use of research findings should also be stated along with the statement of the research problem and the related objectives. To illustrate, different countries use their own National Institutional Ranking Framework to assess the relative positions of the institutions of higher education, say in Engineering and Technology. In some cases, countries go by ranks assigned by some international ranking systems and organizations like the Times Higher Education ranking etc. Any such framework has to specify the criterion variables (unambiguously defined to enable pertinent data to be available for purposes of comparison) and the corresponding weights. This remains a big research problem which should specify whether the relative positions will be linked to government grants for infrastructure improvement or for research promotion or for allowing institutions to go for inter-country collaboration etc.

3

Research Design

3.1 Introduction

Developing a research design keeping in view the objectives of research and the expected outcome and subsequently implementing the design and modifying the same during the course of research, whenever found necessary, are the most crucial aspects of a research process. The research design should provide a complete and yet flexible framework for different activities involved in the research process.

The research design follows from the scope and objective(s) as reflected in the expected outcomes, takes into account limitations on resources available and constraints on their deployment, recognizes special features of the research study like the study being the first of its kind or being mandated to corroborate or reject some previous research findings with a heavy impact on stakeholders, delineates different stages and steps in executing the research programme, gives out procedures to be followed for establishing the validity and generalizability of the research findings and rolls out a comprehensive guideline for various activities to be performed.

It should specify the variables on which data are to be collected including their operational definitions to be used, the choice of proxies or substitutes for variables which cannot be conveniently measured directly, the quantifiable considerations that accommodate latent variables like user-friendliness or convenience in inspection and maintenance of the research output in the form of a new process or product, the manner in which evidences gathered should be analysed, the nature and volume of data to be gathered along with the quality checks to be carried out on such data, the model(s) to be adopted to enable application of different (quantitative) techniques, so that inferences related to the research objective(s) can be reached, and even the way results of research should be validated before being used as inputs to some 'development' effort.

Experiments are almost inevitable in any research, and we know that experiments in many situations proceed sequentially. To achieve the desired degree of precision in the estimates derived from responses in an experiment, we need replications, and replications imply costs. Similarly, in researches that call for planning and conduct of a sample survey to collect information or raw data from some selected units or individuals, we have to identify the sampling units (which could be collections or groups by themselves) – maybe in various stages in some large-scale surveys – and also the responses we have to seek from the units.

The research design is meant to answer all such questions, including those like:

> How many replications do we need? Even before that, we have to decide on the factors to be varied (controlled at several levels) and the number of levels of each factor. We

also have to take a stand on the number and nature of the response variable(s). What constraints are put on factors and responses by ethical issues?

How do we handle multi-response situations and locate the optimum level combination(s) taking care of all the responses? Results flow in sequentially and we should not wait till we have the targeted number of replications. We should look at the responses as they arrive and can even decide when to stop the experiment, as being not likely to provide the desired information or providing the same even before the planned number of replications.

Models of various kinds have to be used to analyse experimental or survey results. Once we know the mechanism of data generation and have the data at hand, how do we choose an appropriate model (from a known basket of models) or develop an appropriate model to represent the data features of interest to us?

We are often motivated to use a simulation model for a real-life process or system or phenomenon to generate relevant data for subsequent analysis and inferencing.

How do we validate a chosen model?

How do we use a model to derive the information or reach the conclusion?

How do we assess the robustness of the information or the conclusion against likely changes in the underlying model?

How do we identify different sources of uncertainty in our data? How do we assess the inferential uncertainty in the inductive inferences we make most of the time?

We should note that inductive inferences we make on the basis of evidences that bear on a phenomenon are not infallible. With adequate evidences of high quality we can provide a strong support to the conclusion we reach. We cannot claim the conclusion to be necessarily true. And if our evidences as part of the premises are weak in terms of quantity or/and quality, the conclusion we reach on analysis of the evidences will involve a lot of uncertainty.

The role of planning or designing a research is of great significance in influencing the research outcome. It is true that unlike in the case of a manufacturing set-up, conformance to design or plan is not that essential in the context of research. In fact, as research proceeds it may be sometimes desirable or even compelling to deviate from some details specified by the initial design. Research, thus, involves a lot of learning through wilful deviation from the path chalked out in the research design. It may even come out that following an initial plan we end up in a failure and learning from such failures is a must. In such cases, the need to leave aside the initial plan and to tread on some new and uncharted ground can well provide inputs for design modification. However, arbitrary or subjective deviations from the plan will definitely affect the quality of output and, thus, of research itself.

3.2 Choice of Variables

It will not be completely wrong to say that any research is essentially concerned with the phenomenon of 'variation' in some features or characteristics or properties or parameters of interest across units or individuals or over time or space or over groups or populations. Phenomena like growth or decline, dispersal or concentration, homogeneity or heterogeneity are essentially variations. Such variations in the variables of interest need observations and subsequent analysis of variables which vary concomitantly with the variables of interest and also those which influence or determine or regulate the variables of interest. In this context, the choice of variables is of paramount importance in any research. This is more easily understood in the case of empirical research. However, this choice is equally important in theoretical studies as well. Thus in a research to develop a more robust procedure or a more efficient algorithm, we need to choose appropriate measures of efficiency or of robustness which again are basically variables with values depending on the deviations from the underlying model that are likely to come up during implementation of the procedure or values in terms of number of iterations.

In the context of empirical researches, relevant variables usually do not admit of unique definitions and unless we go by some defensible and accepted definition of any such variable consistently, our findings will lack validity. Consider a study on the employment–education relationship. People talk loudly about a mismatch between the two. To examine this contentious issue, one needs to develop and use an operational definition of 'employment' and of 'education'. The question of a reference period does automatically arise. Thus to categorize an individual in the age-group 15+ as employed, we usually distinguish between a principal and a subsidiary status in the case of one year as the reference period, And the individual is treated as employed if the individual is engaged in some economic activity, with or without any remuneration, for at least one hour a day on at least 183 days, not necessarily consecutive. And economic activity has not been uniquely defined. It could be some productive activity where the entire produce is consumed by the household itself, without any part going to the market. Of course, the International Conference of Labour Statisticians in its 19th Conference clearly stated that employment work must be work for consumption by others and against pay or profit. Whichever definition is chosen in a given research, it must be consistently adhered to and when results are found to differ from those of others on the same issue, we have to check whether the operational definition of 'employment' has been the same in both the studies. Similarly, coming to education, one category of people will be called literate (not necessarily schooled) if they can read and write a simple sentence with understanding (of what has been written or said). Another definition of literacy could be simply ability to sign. To talk of education, we usually categorize the respondents in terms of the highest level of education received by them. And such levels are not standardized. There could be respondents who have completed the same number of years in formal education, but one gets a graduate-level degree in engineering or medicine while another gets a post-graduate degree in journalism. It is not difficult to assert that different definitions used in different places within the same study will lead to inconsistency of the results. And if non-standard definitions have been used –willingly or unconsciously – the results will not be comparable with similar results produced in other investigations.

In many investigations like customer satisfaction analysis, we must note that 'satisfaction' as a variable from one customer to another (in terms of a degree or just a yes-no category) is an inherent perception and cannot be directly observed. In fact, this is a latent variable, as opposed to 'number of times the customer purchased products of the same brand' or 'number of persons to whom the brand was recommended by the customer' which can be directly observed and is a manifest variable. The level or degree of a latent variable like 'perceived product quality' and not quality as is stamped by the manufacturer or even by a certifying independent agency has to be indirectly derived from some indicator variables like responses to certain questions so framed as to reveal the customer's perception about quality of the product. And a choice of the questions is a matter to be settled in the research design.

3.3 Choice of Proxy Variables

Dimensions of research have expanded enormously with researchers intending, and sometimes keen, to answer difficult questions about the economy, society and the environment. In their quest for more and more information and knowledge, they also embark on cross-country comparisons in regard to indicators of development like education, health, income, communication, energy generation and consumption and the like, and they try to lay hands on credible data bearing on these indicators. Often, they have to depend on official data collected by some government agencies. Occasionally, data they would like to use are not available in the official publications and they have to use data compiled – not always collected through surveys – by private players like industry associations or research agencies etc.

Even when different sources of data are tapped, certain entities like 'R&D intensity' meaning intensity of R&D activities or 'innovative output' implying possibly outputs from innovation or availability of research personnel (professionals), 'use of environment-friendly technologies in manufacturing' and the like, researchers tend to use 'proxies' on which data are available, at least at present or can be conveniently collected. Thus expenditure on R&D is taken as a proxy for 'R&D intensity', sometimes as an absolute figure and some other times as a percentage of total revenue expenditure or total cost of manufacturing or some similar deflator. A problem with the expenditure figure is that the type of activities whose costs are gathered under this head is not the same in all economic units, and in many cases this head of expenditure includes costs of routine activities like inspection and testing of input materials and components.

The World Bank gives out figures for researchers in various countries and relates these figures to the respective population figures. As can be easily understood, the World Bank has to get these figures from the respective countries, which have different statistical reporting systems. Sometimes 'researchers' are taken to be 'professionals engaged full-time in research in industry, academia and laboratories'. In the university system in many countries, faculty members are mostly engaged in teaching and are part-time research workers, while there could be research scholars who are expected to be full-time workers. The number of researchers per million population is a crude summary measure that fails to take into account the age-pyramid for the population, the distribution of researchers according to broad disciplines (not ignoring humanities or social sciences or even inter-disciplinary areas) and the focus on applicability or application in research activities and similar other factors. If such a figure is to be used at all as a correlate or as a determinant of economic prosperity or growth or growth in high-technology exports, one should possibly consider the number of R&D workers engaged in industries only, since those engaged in laboratories are not always working on a mission mode to help industry.

How scientific is it to relate the number of researchers per million persons in a country to national income or share of manufacturing in GDP or export of high-tech products or any such indicator of economic growth? Equally suspect is the attempt to relate the performance of manufacturing industry with R&D expenditure. We come across instances galore where innovations as concrete entities – hard or soft goods – ready to enter the market and impact the economy could be developed without very expensive, time-consuming exercises by large teams of researchers.

While conformity to relevant standards may beget some benefits to the industry, and conformity in many cases is certified by some duly accredited agencies, it must be recognized that certification (for conformity) may be a poor proxy for the set of improvement actions taken by an industry that really lead to improvements in performance. Added to this is the confounding of such benefits also resulting from improvement measures implemented by the industry of itself in areas not covered by the standards.

Genuine data gaps hinder empirical investigations into complex but useful relations needed to describe, predict and control many social and economic phenomena. However, such gaps can hardly justify unimaginative uses of proxy variables which are – more often than not – subjectively chosen and are not closely related to the original variables – manifest or latent – for which such proxies have been used.

Given that we have to use proxies for many latent variables or concept-related parameters, we should now consider the issue of validity of any proxy for a parameter which is difficult or, in some cases, impossible to measure uniquely. We are back to square one. There is no unique definition or measure of validity. Definitions and measures commonly used include predictive validity, concurrent validity, face validity, construct validity and content validity.

Face validity of a proxy variable should be judged in terms of the domain knowledge regarding the original variable and its involvement in the phenomenon being investigated.

To judge predictive validity of a proxy when used to predict the performance or outcome of a suitably defined entity is not a big problem, provided we have a good measure of the outcome and we like to predict the outcome by using the proxy as the only explaining variable. Used along with other variables, the validity of the proxy may be indicated to some extent by the regression coefficient of the proxy in a regression of the outcome on all the explaining variables including the proxy as one.

Similarly, one can assess the concurrent validity of a given proxy by examining the prediction of an outcome based on this given proxy with that based on some other proxy. This exercise is more recommended before we propose to introduce a new proxy in a situation where an existing proxy variable has worked satisfactorily. In situations where more than one proxy variable is indeed required to reflect the impact of the original latent variable on the outcome, we can even examine the contributions of each proxy and choose, if a choice has to be made at all, the one that makes a larger contribution.

3.4 Design for Gathering Data

3.4.1 Need for Data

Data – generically defined – are essential ingredients in any research, be it empirical or theoretical. We need data to study any phenomenon, to reach any conclusion or to arrive at any decision. 'Data' are whatever are given (to the researcher) for any of these tasks, by way of measurements or counts or results of computations or observations directly made or reports and other documents, images and maps, etc. as are being used by the researcher to prove his (her) contention(s). Of course, by data we generally mean numerical or categorical data. The quantity and quality of evidences (loosely taken as 'data') determine the credibility and generalizability of inferences based on the evidences. Thus sample size and the data collection mechanism play a significant role in all empirical research. Data generated by others could be collected by the research worker or could be generated by the research worker himself/herself.

Evidences to provide support to the research hypotheses eventually leading to the acceptance or negation of any hypothesis have to be collected in any research activity – theoretical or empirical. We generally take data as evidence. It must be remembered that the Greek word 'data' implies 'whatever are given' to the research worker to study the underlying phenomena and/or to make relevant inferences. Besides numerical data pertaining to individuals or units covered by the underlying phenomenon, data may include signatures or graphs or maps or documents or similar other sources of information. Data which do not throw any light on the phenomenon under study do not qualify to be called evidence. In subsequent discussions, we take for granted that data are evidences and form part of the premises in any inference exercise.

We tend to feel that there is no role of data collection or analysis in theoretical research. This is definitely not true. In the context of theoretical research offering a new derivation of an existing result or coming up with a new result or proposing a new method or technique of data analysis or even new software, data would have to be generated to establish the novelty or generality or simplicity or efficiency in terms of suitable criteria (known or suggested in the research) whose values have to be calculated on relevant data already quoted in the literature or generated through simulation or even on primary data conveniently collected by the research worker.

3.4.2 Mechanisms for Data Collection

Depending on the objective(s), research may require primary data to be generated by the research worker, secondary data to be gathered from relevant existing sources with or without editing, or both. To generate primary data pertaining to the underlying system or process or phenomenon, we may have to design and conduct an experiment or plan and conduct a survey, most often on a sample of experimental units. Sometimes, we gather secondary data from several different sources. Before we pool such data, we must make sure that they are 'poolable' in terms of their purposes, operational definitions of various terms and phrases, collection and verification mechanisms, and the like.

3.4.3 Design for Data Collection

The data collection exercise has to be both effective and efficient and, for this reason, research design has to involve the right design for data collection whether through experiments or through surveys. The design has to be effective in the sense of providing the right type of data in the right quantity and of the right quality to throw adequate light on the phenomenon or system under investigation and to allow appropriate inferences to be reached, including statistical tests of significance whenever needed. It has to be efficient in the sense that the use of resources for the above purpose has to be kept at the minimum possible.

Primary data have to be generated, either by conducting an experiment in a laboratory or in the field which has to be properly designed and conducted to yield the desired data or by planning and conducting a sample survey which has to be properly planned. Usually sample surveys are conducted with an appropriate sampling design. The basic difference between these two data-gathering exercises is that units or individuals in a survey are observed or assessed as they are, with no intervention on the units, while these are subjected to some treatment in an experiment to provide some responses which are noted in respect of the units or individuals treated.

Since the ambit and diversity of research these days is enormous, experiments and surveys have also to be generically and widely defined and delineated to accommodate data needs for various types of research with distinct objectives. Thus, an experiment generically means an exercise to gather empirical evidence (not based on theory) and acquire knowledge relating to a phenomenon (happening in nature or the economy or society) or a system or an operation. Experiments range from simple and quick observations, through design, development and administration of some treatment, to complicated and even hazardous ventures into unknown tracts and entities. Simple laboratory experiments to identify and estimate the properties of substances, bioassays to determine the potency of a new biological preparation compared to a standard one, agricultural field trials with different manures, fertilizers and irrigation practices to study their impacts on some crop, clinical trials with different drugs or treatment protocols, experiments to develop appropriate specifications for a manufacturing process like synthesis of an organic molecule, complicated and costly experiments in space research, etc., reveal the panorama of experiments.

Similar is the case with sample surveys. We can think of a survey to estimate a single parameter like the labour force participation rate from a household-based survey in a particular district. Attempting to cover the entire country will mandate second-stage stratification of households within each village or town selected in the first stage from the district/state. In a survey to estimate the number and demographic features of substance abusers in the country, we cannot consider an abuser as a sampling unit since a sampling frame does not exist. We may have to identify places where groups of such people generally meet and follow up leads that we may get to secure a sample that may not provide an 'unbiased' estimate. In a survey to assess drop-out rates in primary schools, we should trace a group of children enrolled in class 1 throughout higher classes till, say, class IV.

While in most surveys covering individuals or households or even establishments, we collect information from some informant or respondent (who may not be the ultimate sampling unit), a research on crop diversity or animal diversity in a certain region may not need an informant, The investigator does it all. As an alternative to an opinion survey on a crucial national issue raging for quite some time, where a representative sample of informed people has to be covered, we can think of a survey where the sampling units are newspaper editorials on the issue published in different newspapers with a minimum specified readership carefully analysed by some experts.

A research worker has to first identify the type of data needed and the way the data should be collected.

Experiments are conducted to

- Explore new relations connecting various entities

    (*objects* and *processes* involving them along with the *outcomes* of such processes)
- Establish new properties and new functions of existing materials or products or services
- Develop specifications for various parameters of the final product and, going back, for material and process parameters
- Demonstrate success or failure of a new product design in realizing the 'mission' behind the creation of the product

We can reasonably look upon an experiment as an effort to produce an outcome that will throw some light on the relation connecting some response variable(s) with some factor(s) which can be controlled at given levels.

Factors in a laboratory experiment could be

- Process variables (process/equipment/environment parameters) like reaction time or temperature, proportions in which some ingredients are mixed, etc.
- Input variables (input parameters) like molar ratio, pH level of solvent, etc.

Response(s) could be yield, purity or concentration, residual toxicity, taste or aroma, gain in weight, increase in hardness, etc.

Factors are controlled at given levels/doses/values/types by the experimenter. These levels can be purposively varied from one run of the experiment to another or even from one experimental unit to a second within the same run to examine corresponding variations in responses Thus, factors are non-random variables.

Properties or features or changes in such properties or features which follow the application of a treatment or treatment combination on each experimental unit which are ultimately analysed to meet the objective(s) of the experiment and which are expected to vary from one unit to another are recognized as response variables. Most experiments involve only one response variable, though multi-response experiments may well be required in some research or other.

Factors are wilfully allowed to vary, responses vary as a consequence. Responses vary over units or runs, with or without variations in levels of factors. And such variations are due to a number of causes not all known to us and which cannot be exactly predicted. Therefore, responses are random variables. Factors remaining at the same levels lead to unexplained or random variations in responses. When such levels vary, they cause assignable variations in responses.

Factors may be called *predictor variables*, while responses correspond to values of the *predictand variable(s)*.

Given this, the objectives of an experiment are to

- Identify factors (out of the whole set of postulated or reported factors) which have significant effects on the response(s) and should be screened for subsequent study
- Determine the effect of a causal entity on the response it produces
- Develop a suitable model or a functional relation between the response(s) and a set of factors (predictor variables)
- Use such a model (usually represented by a response curve or surface) to adjust the factor levels that lead to improved responses and finally to yield the 'best' combination of factor levels
- Thus screening the possible factors, developing a factor-response model and determining the optimal factor levels are the three major objectives of an experiment

Most experiments are meant to compare the effects of two or more treatments (factors or levels of a factor or factor-level combinations).

Some experiments, as in a bio-assay, are aimed at finding levels of two different factors (a standard and a test biological preparation in the case of a direct assay) which produce the same effect.

In fact, data required to (1) estimate some unknown parameter like the potency of some biological preparation relative to a standard one or (2) test a hypothesis about effects of two different media for quenching of an alloy on its hardness or (3) take a decision on the temperature to be maintained in a calcination plant to reduce sulphur content in crude petroleum coke are to be generated through duly conducted experiments which can be branded as controlled sampling experiments. While the first two problems require comparative experiments, the third calls for an optimization experiment.

While research has always involved some trial and error – possibly more in earlier days than at present – resource limitations make it almost mandatory that even such trials be so planned that they can lead to the desired result(s) as quickly as possible. For example in an experiment to find out the composition of a metallic alloy in terms of proportions of basic metals and non-metallic substances in such a manner that a structure made out of the alloy will have the longest time to rupture under some specified stress situation should use a sequential design for experimentation, rather than any fractional factorial design or any other design attempting to minimize the number of design points or of experimental runs. An experiment in this case is quite time-consuming even if we make use of accelerated life testing methods and therefore the need for 'optimization' is much stronger than in some other situations. The objective of research could even require the experiment to achieve the longest time to rupture subject to a minimum desired ductility. This will complicate the analysis of experimental results.

Most experiments involve multiple factors which are controlled at several levels each, all possible combinations of factor levels or some selected combinations are applied to more than one experimental units each and each experimental unit yields some response(s). The basic principles of randomization, replication and local control are observed to ensure effectiveness. It may be pointed out that all these three principles may not be applicable in all situations.

Multi-factor or factorial experiments provide a methodology for studying the inter-relations among multiple factors of interest to the experimenter. These experiments are far more efficient and effective than the intuitively appealing but faulty one-at-a-time experiments. Factorial experiments are particularly well-suited for determining that a factor behaves differently (as reflected in the experimental response) in the presence, absence or at other levels of other factors. Frequently, a breakthrough in quality comes from the synergy revealed in a study of interactions. If the number of factors happens to be large, then a fractional factorial design offers a possible compromise.

Dealing with industrial research and industrial experiments in that connection, we should note that an experimental unit in this case is one run of the experiment allowing some controllable factors to be at pre-specified levels and using an equipment that is required for each run. The run takes some completion time and yields a response or several responses. For example, a pharmaceutical experiment may have to be conducted in a reactor vessel where in each run we choose one of two possible catalysts, proportions of some basic chemicals to be mixed up, the temperature and the time for reaction. And two responses, viz. yield and purity or concentration, are of interest to the researcher. In the case of such an experiment, the principle of randomization makes sense only if there is any residual effect at the end of one run on the responses for the next run. And the principle of local control is really not relevant.

Design requirements to meet effectiveness and efficiency criteria apply strictly to experiments being currently planned to be conducted hereinafter. There are situations in real life where some sort of an experiment was initiated earlier and may be continuing now, which can be looked upon somewhat indirectly as multifactor experiments with the objective of choosing the optimal treatment or factor-level combination. This is like a retrospective experiment against the prospective experiments for which different designs are talked of. A noticeable element missing in this context will be randomization, which is basic for applying any statistical methods for making valid inferences.

Experiments involved in scientific research may differ in their objectives and accordingly justify different designs. An experiment to determine the tensile strength of yarn coming out of a spinning mill may not even require a design, except the choice of the test method, the measuring equipment., the control over environmental conditions, etc. In fact, such experiments often characterized as 'absolute' experiments are quite commonly involved in physical sciences. An absolute experiment conducted to yield the value of a measurand for a physical object or a chemical solution or a biological preparation or the distance of a star from the Earth may require the setting up of an experiment that will provide multiple measurements to be appropriately combined to give the desired value. In such a case, the experimenter has to be careful about the choice of measurements to be produced and the way these measurements have to be combined.

On the other hand, if in the tensile-strength measuring experiment one is required to find out machine difference in strength, the design must incorporate the replication consideration and control over factors other than the influence of machine on strength. If the experimenter wants to study the effects of batching producing slivers, machines and operators, a three-factor experiment has to be designed incorporating all the three basic principles mentioned earlier. In these cases, yarn strength is the only response variable to be analysed. However, there could be situations in which more than one response variable could be involved.

Depending on the objective, we should distinguish among screening experiments to find out which of the factors suggested from domain knowledge are really significant in terms of their effects on the response variable(s), comparative experiments which tell us which treatment (factor-level combination) is better than others and optimization experiments which are meant to locate the treatment that results in the optimum (maximum or minimum) response by studying the response surface.

While designs for different purposes in different situations have been separately discussed in a subsequent chapter, we simply point out here that the choice of levels in respect of factors recommended on the basis of a screening experiment or otherwise is itself an important decision. To keep costs and hence the number of design points or factor-level combinations under control and, at the same time, to make inferences based on the experiment to remain valid over a the range of levels of interest for each factor, we have to choose the few levels of each factor in such a way that the corresponding factors can be kept controlled at the chosen levels in terms of available equipment. We should also check compatibility condition for each factor-level combination. In practice, it is possible that some level of one factor A is not compatible with some level of another factor B. Similarly important is the choice of the response variable(s), which could be categorical or numerical. In the case of a categorical variable, the categories have to be unequivocally specified. For a measurable response, measurement accuracy and consistency have to be ensured.

As part of a research undertaken by the human resource development department in a service organization, the research team wanted to find the best way of conducting workshops / seminars / training programmes for its executives at senior and middle levels in both hard and soft areas of operation. The department has been organizing such programmes by involving both internal seniors and external experts as faculty members, sometimes for half a day and sometimes for a full day or even for a couple of days. In some cases, the programmes started or were held on Mondays or Fridays, while in some other cases it was any weekday. To judge effectiveness, a feed-back form would be filled up by the participants by the end of a programme. In this context, the need was felt for a factorial design involving the following factors and their levels:

Faculty: three levels, viz. internal wholly, external wholly, a mix of the two

Duration: half-day, one day, two consecutive days, two days with a gap

Day / starting day: Monday, Friday, any other weekday

Participation: only senior (mixed) level executives, mix of the two

The perceived overall effectiveness averaged over the participants was the response variable. There was no problem in accessing data for some programmes already held, and a few more being planned were also included, each programme corresponding to an experimental unit. Feedback about the revealed impact of participation in a programme on performance in a related task to be obtained after some time gap could have been a better response variable. That, of course, is a different research question.

This experiment, partly prospective and partly retrospective, lacks the elements of randomization and to some extent local control expected in a full factorial design. And such experiments (some call such exercises quasi-experiments) are not that uncommon in practice. We must remember that randomization is a key assumption to justify statistical analysis of the data for inference purposes.

A sampling design for a survey (usually a field survey) involves a number of decisions prior to the conduct of the survey to collect data. We need to define the sampling unit, each unit to yield some response(s), the sampling frame or the totality of sampling units from which a representative sample will be selected for the survey at hand, the parameter(s) of interest which have to be estimated and the level of accuracy (relative standard error) to be achieved. Thereafter we decide on the sample size, the method of selecting units in the sample and the method of estimation.

Quite often we use multi-stage stratified sampling in sampling from a large population which is likely to be heterogeneous in respect of the variable(s) of interest but may admit of relatively homogeneous groups or clusters or strata. In such cases, we might include a sub-sample from each cluster in the overall sample. In some cases. such clusters can be easily identified as natural groups corresponding to geographical or administrative divisions or social groups and the like. There could arise situations where strata have to be developed by identifying some appropriate stratification variable correlated with the estimation variable of interest, and determining suitable cut-off points on its range to yield the strata, having chosen the number of strata earlier.

In practice, research workers face problems like non-availability of a sampling frame in the sense of a complete list of sampling units as are of direct interest in the research, possible avoidance of truthful answers to confidential or private questions, inaccessibility of certain segments of the population, and a host of others. Thus in the case of a sample survey to estimate the total number of drug abusers in a state or a country, the population of such individuals is not defined and such individuals are not evenly distributed across different parts of the state or country. Further, individuals selected for the survey may not provide a truthful response to a question regarding possible drug abuse. In such a case, we need to adopt some design like chain sampling with linear or exponential referral and the randomized response technique.

In certain surveys, the focus is on likely changes with respect to some variable(s) of interest over time taking place in a group of units. We can adopt a cohort sampling design or a panel design. The panel to be followed up could have the same units throughout, or we can allow a portion of the panel to be replaced by a sub-panel of new units. This is sometimes referred to as a rotational panel design. Thus one may be interested to track the employment situation or some performance parameter like gross value addition in a panel of industrial establishments over successive years.

Different sampling designs which suit specific objectives and for which standard methods for parameter estimation are available are discussed in chapter 4 on data collection.

3.5 Measurement Design

Data from a planned survey will come up as observations or counts or measurements on variable features possessed by different (ultimate) sampling units covered by the survey. In a designed experiment, we have data on (a) treatment or combination of factor-levels applied to each experimental unit along with value(s) or level(s) of any co-variate(s) or con-ditioning variable(s), if any, and (b) response(s) from the experimental unit. The variable(s) to be noted in a sample survey and the response(s) to be noted in an experiment must be well-defined and amenable to proper categorization or counts or measurements. In the case of an experiment, the factors (whose levels will be recorded) must be such that they can be controlled at desired levels.

Good laboratory practice is a mandatory requirement to carry out experiments, to record measurements of factors and responses, to carry out checks for repeatability and reprodu-cibility of test methods, to estimate uncertainty in measurements, to get back to original records in case of any doubt, and thus to provide a credible basis for analysis of results Even when we are considering a field experiment, responses –usually called 'yields' – have to be assessed for accuracy and precision. Results of such an analysis should be duly interpreted in the context of the experiment and the objective behind the same.

The result of measuring a property of a physical entity may depend on the value(s) of some parameter(s) assumed in the experiment, and any inaccuracy or uncertainty in the value(s) of any such parameter(s) will be reflected in the quality of the resulting measure-ment. Feynman (1992) pointed out that Millikan and Fletcher had measured the charge on an electron by an experiment with falling oil droplets and got an answer which was not really correct, because they took the incorrect value of viscosity of air. (Viscosity of air, like that of any other fluid or fluid mixture, can be kinetic or dynamic and does depend on ambient temperature.) In fact, all subsequent results on the charge came out to be larger, till finally this deficiency was detected. Millikan and Fletcher obtained a value 1.602×10^{-19} C while the currently accepted value stands at 1.592×10^{-19} C. Since the result e is a funda-mental physical constant, its value has to be precisely known. Incidentally, this Nobel Prize-winning experiment was quite an imaginative one, though not so complicated or elaborate.

Since the quality of measurements is of great significance in experimental sciences as also in engineering and technology, it will be quite in order to briefly discuss the various concepts, measures and practices in this connection. And we attempt this in the next section.

3.6 Quality of Measurements

Measurements play a significant role in all scientific and technological investigations and such measurements are derived by different operators working on different instruments under different controlled environmental set-ups. Validity and generalizability of research findings are circumscribed by the quality of measurements, judged in terms of accuracy and precision. These two measures are better understood in terms of repeat measurements on the same property or feature of the object under consideration (usually called the measurand and denoted as X). We take for granted that each measurand has a true value T which is indeterminate, except when known in theory. Let us consider repeat measurements $x_1, x_2, \ldots x_n$ made in the same laboratory using the same instrument by the same operator under the same controlled environmental conditions. Let m be the mean of these measurements and s the standard deviation. Then accuracy of the measurements is inversely estimated by $\mid m - T \mid$ and precision inversely by s. Since T is not known, it is usually replaced by an accepted reference value for T.

Current interest lies in estimating the uncertainty about the true value of the measurand (a feature or characteristic to be measured) based on a single observed value or, at best, a set of repeat measurements. The uncertainty in measurement is a parameter, associated with the result of a measurement, that characterizes the dispersion of the true values which could reasonably be attributed to the measurand. The parameter may be, for example, the standard deviation (or a given multiple of it), or the half-width of an interval having a stated level of confidence.

Measurands are particular quantities subject to measurement. One usually deals with only one measurand or output quantity Y that depends upon a number of input quantities X_i ($i = 1, 2, \ldots, N$) according to the functional relationship.

$$Y = f(X_1, X_2, \ldots, X_N) \tag{3.5.1}$$

The model function f represents the procedure of measurement and the method of evaluation. It describes how values of Y are obtained from values of X_i.

In most cases it will be an analytical expression, but it may also be an analytical expression which includes corrections and correction factors for systematic effects, thereby leading to a more complicated relationship that is now written down as one function explicitly. Further, f may be determined experimentally, or exist only as a computer algorithm that must be evaluated numerically, or it may be a combination of all these.

An estimate of the measurand Y (output estimate), denoted by y, is obtained from Eq. (3.5.1) using input estimates x_i for the values of the input quantities X_i.

$$y = f(x_1, x_2, \ldots, x_n) \tag{3.5.2}$$

It is understood that the input values are best estimates that have been corrected for all effects significant for the model. If not, necessary corrections have been introduced as separate input quantities.

Standard uncertainty in measurement associated with the output estimate y, denoted by u(y), is the standard deviation of the unknown (true) values of the measurand Y corresponding to the output estimate y. It is to be determined from the model Eq. (3.5.1) using estimates x_i of the input quantities X_i and their associated standard uncertainties $u(x_i)$.

The set of input X_i may be grouped into two categories according to the way in which the value of the quantity and its associated uncertainty have been determined.

Quantities whose estimate and associated uncertainty are directly determined in the current measurement. These values may be obtained, for example, from a single observation, repeated observations, or judgement based on experience. They may involve the determination of corrections to instrument readings as well as corrections for influence quantities, such as ambient temperature, barometric pressure or humidity.

Quantities whose estimate and associated uncertainty are brought into the measurement from external sources, such as quantities associated with calibrated measurement standards, certified reference materials or reference data obtained from handbooks.

The standard uncertainty in the result of a measurement, when that result is obtained from the values of a number of other quantities, is termed combined standard uncertainty.

An expanded uncertainty is obtained by multiplying the combined standard uncertainty by a coverage factor. This, in essence, yields an interval that is likely to cover the true value of the measurand with a stated high level of confidence.

The standard uncertainty of Y is given by

$$\sigma_y = \left\{ \sum \sum C_i C_j \sigma_{ij} \right\}^{1/2} \tag{3.5.3}$$

where inputs X_i and X_j have a covariance σ_{ij} and $C_i = (\delta f / \delta X_i)$ at $X_i = x_i$ is the sensitivity of Y with respect to variation in X_i. The formula simplifies if the inputs are uncorrelated. The variances can then be easily estimated if repeat measurements are available on an input; otherwise these are estimated by assuming some distribution of true values (which could be made to correspond to the same observed value) e.g. normal or rectangular or (right) triangular etc.

The uncertainty analysis of a measurement – sometimes called an uncertainty budget – should include a list of all sources of uncertainty together with the associated standard uncertainties of measurement and the methods for evaluating them. For repeated measurements the number n of observations also has to be stated. For the sake of clarity, it is recommended to present the data relevant to this analysis in the form of a table. In this table, all quantities are to be referred to by a physical symbol X or a short identifier. For each of them at least the estimate of x, the associated standard uncertainty of measurement U(x), the sensitivity coefficient c and the different uncertainty contributions to u(y) should be specified. The dimension of each of the quantities should also be stated with the numerical values given in the table.

The concept of standard and expanded uncertainty and the associated caution in interpreting measurements resulting from experiments has to be given due attention, particularly when comparing results obtained on the same measurand on different occasions in different laboratories and following different measurement methods. In fact, this question is quite serious when some non-standard method has been used to measure a parameter at the micro-level. A numerical example of uncertainty estimation follows.

The tensile strength testing machine in a conveyor belt manufacturing unit is calibrated annually. Tensile strength of finished belts is determined using the equipment involving the tensile value disk and the load cell. Ten repeat measurements on tension in kg/cm² were available on a particular belt specimen to estimate uncertainty about the true value. The following information about the equipment was also available for the purpose.

Tensile value disk	
Range used for calibration	0–50 kgf
Accuracy	As per manufacturer's data
Resolution	1 div. = 0.1 kgf
Load cell	
Uncertainty (%) from its calibration certificate	0.37 (A1)

Readings on tension are reproduced below

Reading No.	Tension	Reading No.	Tension
1	153.50	6	160.39
2	159.78	7	187.05
3	167.04	8	156.12
4	161.83	9	161.39
5	156.10	10	160.83

Type B Evaluation

Uncertainty of load cell received from the corresponding calibration certificate. We assume the underlying distribution to be normal so that the coverage factor at 95% confidence level is approximately 2. Thus U_1 (%) = A_1 / 2 = 0.37 / 2 = 0.185%.

(A_1 considered as the expanded uncertainty U_e = 2 × standard uncertainty.) Thus:

Uncertainty of load cell U_1 = 0.185 * 160.40 * 0.01 = 0.297 kg/cm^2.

Since U_l = U_l % mean reading / 100. Thus the estimated uncertainty of the load cell works out as 0.37 * 160.40 * 0.01 = 0.593 kg/cm^2.

Combined standard uncertainty U_c = √ [U_r * Ur + U_l * U_l] = 1.46 kg/cm^2 and % U_c = 0.91% = U_c * 100 / mean reading.

Expanded combined uncertainty for approximately 95% level of confidence

$$U = 2 * 1.46 = 2.92 \text{ kg/cm}^2.$$

And U % = 1.8%.

The uncertainty budget can now be worked out conveniently.

3.7 Design of Analysis

Types of analysis to be done and methods and techniques to be adopted should take due account of the nature of data collected and also the mechanism for data collection, besides the objectives of research. One basic principle to be remembered in connection with data analysis is that objectives of research and data to be analysed determine the nature and extent of sophistication in analysis, but the researcher's knowledge of or access to some tools and techniques of analysis do not determine the objectives and the nature and extent of data to be collected and subsequently analysed.

While using methods and tools for statistical analysis, we have to recognize the fact that application of such methods and tools most often requires certain assumptions to be made, and the assumptions may not be justified by the nature of the data and the way these were collected. Thus, normality of the distribution of some underlying variable, independence of successive observations/measurements, additivity of some effects on the variable exerted by some factors, constancy of variance of the variables involved, and the like are commonly assumed in many inferential tools in statistics. At the same time, we also know that if successive observations on some characteristic of the product are considered, say, from a chemical process, then it is quite likely that several successive observations will be correlated among themselves because of some cause-system operating on the process during some time when the observations were made. In many studies, we may analyse data on a variable known to have a highly skewed distribution like that of income among households in a town or a distribution with a heavy tail. In such cases, assumption of normality may not hold and we may think of a normalizing transformation of the variable. In all such cases, the research worker should first apply suitable tests to examine the validity of the assumptions to be made in the model for analysis. It is true, however, that some methods and tools are sufficiently robust against violations of the assumptions. This leads us to apply distribution-free methods or model-free induction.

In exploratory research which is primarily descriptive or at best analytical in nature, most techniques for analysis of the data do not call for any model. Of course, some techniques involved in processing the data collected like methods of scaling responses to items in a questionnaire by Likert's method may require the assumption of normality for the underlying trait. And some research of this type may attempt to come up with a probability model for some of the variable(s) on which data have been collected and proceed further to put forth some descriptive or analytical measures, say for the purpose of comparison, which are based on the distribution justified by the data.

However, for research offering predictions for the future or ending up in inferences which will be valid even beyond the set of data collected to a wider group (population, say) it is imperative to assume a model and establish its validity and suitability or adequacy before making inferences based on the assumed model.

3.8 Credibility and Generalizability of Findings

An important concern in a research investigation is to ensure that research findings are based on adequate evidence, appropriately analysed and duly interpreted in the context of the objectives of research. In empirical research with evidence gathered through surveys or experiments, results finally reached are expected to remain valid even outside the domain or coverage of the survey or the experiment, unless otherwise required and accordingly stated. Findings of empirical research are usually amenable to generalization beyond the evidence considered. And the experiments should be reproducible across locations or laboratories and over occasions. Similarly, in survey-based research using a particular sampling design and eventually involving a particular sample, results obtained are not expected to be identical with findings that come out of a different sampling design or a different sample following the same design. However, results obtained from different samples should vary within some limits which can be worked out. Similarly, results yielded by experiments with a selected set of factors and their levels should not vary remarkably from one such set to another.

As pointed out in connection with good laboratory practice, test/inspection/measurement results should all be recorded, even when some of these appear to be outliers or apparently inconsistent with others or suspected to be wrong. The practice of 'good results in, bad results out' as was had recourse to by some earlier scientists should not be encouraged. And such tests and measurements must be made by standard methods with known inherent accuracy.

There could be exceptions where a research has a narrowly specified ambit and its findings are not expected to be applicable outside the framework circumscribed by the data set arising from an experiment or a survey. In fact, the need to ensure generalizability may not be a requisite case where the domain of research is small to allow complete coverage and the domain is dynamic in the sense of having features which vary from time to time.

Inadequate sample size, non-trivial non-sampling errors or response biases, high proportions of missing values, inappropriate choice of models or of techniques for analysis, unimaginative use of software, conclusions not relating to the objectives of research, deliberate or motivated manipulations of the original results and similar other factors tend to detract from the credibility of research results.

In the case of survey-based research, we have to estimate some parameter(s) of interest from the sample data. Thus, we may estimate the total number of workers in establishments with nine or fewer workers engaged in manufacturing or the average number of hours worked by casual workers during a week or the proportion of people suffering from a particular disease. Unless some estimated relative standard error is associated with the estimate of a population parameter, it will be difficult to assess the credibility of the estimate. The relative standard error of the sample mean is the same as the relative standard error (RSE) of the sample total and is given by the formula

$$\text{Relative Standard Error} = \sigma / (\mu \sqrt{n}) \sqrt{[(N - n) / (N - 1)]}$$

where n = sample size, N is the population size, σ is the population standard deviation of the variable under study and μ is the population mean. It must be noted that this formula for RSE is derived under the assumption of simple random sampling without replacement. For any other sampling design like a multistage stratified random or systematic sampling an inflation factor called the design effect has to be applied. The RSE expression involves population parameters and an estimated RSE can be obtained by replacing the parameter values by their sample estimates. The higher the relative standard error, the wider is the confidence interval for the parameter being estimated. Usually, a value of RSE not exceeding 0.10 is desirable. The real problem is to find a sample size to keep the RSE at a specified level without a knowledge of the estimates prior to data collection. In the case of surveys which are repeated, ideas about the population parameters could be obtained from the sample estimates worked out in some previous survey round. In most cases, the population size being much larger than the sample size, the finite population correction factor is ignored. Further, if the RSE without the correction factor does not exceed a certain value, the one which takes the correction factor into account will be smaller still. The RSE of the sample estimate p for the parameter π is given by the expression $\sqrt{[\pi / n (1 - \pi)]}$ without the above finite population correction. And in the absence of any prior knowledge about the value of π one can consider the value that maximizes the RSE and get an upper bound to the RSE.

3.9 Interpretation of Results

While a comprehensive description of a subtle and/or complex system that can facilitate its understanding could be the objective of some research, most researches are meant to reach some inference or conclusion regarding some aspect of the system or phenomenon under study. Incidentally, we have two mechanisms for acquiring knowledge, viz. (1) directly by perception through the senses whose power is often greatly enhanced by various instruments and technologies and (2) indirectly, through inferencing or reasoning. There are two ways of reasoning and inferencing, viz. deduction and induction. In fact, both the mechanisms and both types of inferences are involved in most research to acquire knowledge. The way such mechanisms and inferences will be attempted should form a critical part of the research design.

Let us have a look at the problem of 'inferencing' where we start with some premises, go through some intelligent and logical processing of evidence within premises and reach some conclusion or inference. The premises which constitute the building block of the exercise include relevant postulates or propositions or axioms, evidence bearing on the underlying phenomenon, any previous knowledge that may throw some light on the phenomenon, besides the 'Model', which is a set of assumptions made – of course, with adequate justification – to enable processing of the data in the given context. As stated earlier, whatever are given in the premises to enable a conclusion or an inference to be reached should be called 'data'. The data are summarized, analysed and integrated logically, with the help of appropriate methods., techniques and software. And finally we come to conclude or infer.

A fundamental difference between the two types of inferences is that in deduction – which pervades what are branded as exact sciences and almost the whole of Pure Mathematics – the premises are sufficient to warrant the conclusion. Within the premises, there could be some 'core' part which is sufficient to reach the conclusion without needing the remaining part. A deductive inference holds necessarily in all cases and is not associated with any 'uncertainty'. In induction, on the other hand, the premises can only provide some support to the conclusion, but cannot guarantee the latter. 'Some' support may imply relatively strong support in favour of the conclusion or could be relatively weak. There is no 'core' part of the premises and as evidences grow or are modified, the support extended to the conclusion may also change. A simple situation to illustrate is sampling inspection, say by an attribute quality parameter classifying items from a lot as defective and non-defective to weigh the conclusion 'lot is good'. Suppose, initially we come across only one defective out of the first 20 items inspected and as we inspect 15 more items we detect three more defective items. Here, on the basis of the initial data and sample size 20, we may support the conclusion rather strongly, but coming to consider the bigger sample size of 35 items yielding five defectives we tend to withdraw our support. In induction, evidences bearing on the conclusion and collected to reflect on the underlying phenomenon constitute an essential component of the premises. In deduction, the premises could be wholly in terms of some postulates. Thus, 'man is mortal' is a sufficient core premise to warrant the conclusion that Ram (a male) will die, no matter whether Ram is quite aged, suffering from some diseases or has no one to take care of him. In deduction, on the other hand, evidences play the major role and because of possible uncertainties associated with the evidences and with the processing of evidences through methods and techniques that acknowledge and accommodate unpredictable 'chance' elements, inductive inferences invite uncertainties.

There are three distinct situations where we make inductive inferences. The first is when we like to look back at the past (sometimes remote past) and to infer about some phenomenon that took place then on the basis of evidences now available. We go from the known 'present' to the unknown 'past'. In the next case, we tend to make predictions or forecasts about some future phenomenon from evidence available now. Here, the 'future' is unknown and we can only make 'uncertain' inferences about the future. The third and the most common situation warranting inductive inferences is where we like to make inferences about the 'whole' or the 'population' based on evidences that are contained only in a 'part' or a 'sample'. In fact, we will be dealing with this third situation in most of the studies that we plan and conduct.

An induction may be open, the proposition being suggested by the premises of induction without reference to any preconception; or it may be hypothetical starting with some preconceived hypothesis(ses) which should be confirmed or rejected in the light of the premises. In the former case, induction is used as a tool for discovery, in the latter as one for testing. In statistical induction, the evidence represents a collection of contingent facts, its content being never fully known before it is collected. The set of all possible forms that the evidence may take is clearly defined beforehand and is called the domain of evidence.

Such inferences bear on some hypotheses relating to some earlier result about the system or some findings from the present investigation. In this connection, it may be pointed out that a hypothesis is a statement that is neither accepted instantly nor rejected downright. Instead, the statement is subjected to verification on the basis of evidence available before it is rejected or accepted. We thus consider only testable or verifiable hypotheses. It may be added that evidences already gathered may determine whether or not a particular hypothesis of interest can be verified or not. This points to the need for collection of evidences that will be required to verify the (working) hypotheses formulated as part of the research objectives.

3.10 Testing Statistical Hypotheses

A statistical hypothesis is a statement relating to some feature or features of the underlying phenomenon and the corresponding patterns of variation revealed by the different variables involved, about relations connecting the different variables, about possible differences between two or more parameter values, and the like. The initial statement about any such aspect of the problem which is set up for possible validation is usually referred to as the 'null hypothesis'. In fact, developing appropriate null hypotheses reflecting the expected outcomes of the current research is a critical step in the research process. Sometimes, we may set up quite a few null hypotheses at the beginning and, as we proceed with collecting evidence and/or acquiring more and more information about the initial objectives, we may drop some of the initial null hypotheses or even modify some of those. This is why expected outcomes have sometimes been identified with 'working hypotheses'.

While detailed analyses of data have to be taken up only after the intended data have been collected and peculiarities in data like incompleteness, missing-ness, dubious values, false responses have been noted, research design should give out broad hints about the types of analysis as well as the manner in which hypotheses will be verified and results of verification exercises recorded. In the context of testing statistical hypotheses, it may be noted that the formulation of 'alternative' hypotheses – one-sided or two-sided – is not mandated in all research investigations. We may simply remain interested in the 'null' hypotheses that we have framed and like to know if the 'sample' of evidences that we have is strong enough against the concerned null hypothesis or not. Thus either we are able to reject the null hypothesis or we fail to do so. In fact, the p-value advocated by Fisher is a (negative) measure of the strength of evidence against the null hypothesis contained in the sample. This is obtained as the probability P_0 that a sample from the underlying population(s) will yield a value of the test statistic that will exceed the value obtained from the data on the assumption that the null hypothesis is true. We require the null distribution (distribution under the null hypothesis) of the test statistic which is a function of the sample observations and takes into account the null hypothesis to be verified. In fact, the test statistic T (x) may be looked upon as a measure of discrepancy between the null hypothesis and the sample observation (x), with a larger value implying a greater discrepancy. The probability of T exceeding the observed value, say $t_0 = T(x_0)$, is called the level of significance or tail probability or simply p-value. Usually, the observed value t_0 is taken as significant or highly significant according to whether $P_0 < 0.05$ or < 0.01. Fisher suggested that values of P be interpreted as a sort of "rational and well-defined measure of reluctance to the acceptance of hypotheses".

In the above approach, several questions have to be kept in mind. The first concerns the choice of a test statistic for which no general theory exists. However, the likelihood ratio can serve the desiderata to a large extent. The second is the fixity of the P value to a fixed sampling experiment with a given sample size and also a given model.

However, the formulation of narrowly or somewhat broadly defined alternative hypotheses may also be quite warranted in some other research studies where we need to evaluate a posteriori which hypotheses – null or alternative – provide a stronger support to the observed sample as being more likely to be observed. In such situations, we develop and use the concepts of Type I and Type II errors associated with each test procedure and we try to fix the probability of committing the Type I error at a pre-specified low level like 0.01 or 0.05 and to minimize the other probability of not rejecting the null hypothesis when it is false in reality. The theory of most powerful tests was introduced by Neyman and Pearson.

Fisher, Jeffreys and Neyman disagreed on the basic numbers to be reported in testing a statistical hypothesis and considerably different conclusions may be motivated by these different numbers. Berger (2003) puts up an illustration where the data $X_1, X_2, \ldots X_n$ are independent and identically distributed (IID) from the normal (μ, σ^2) distribution, with known σ, and n =10, and it is desired to test H_0: $\mu = 0$ versus H_1: $\mu \neq 0$. Also, suppose z= $\sqrt{n}$ m / σ 2.3 (or s = 2.9).

Fisher would report the p-values p = 0.021 (or p = 0.0037). Thus H_0 would be rejected.

Jeffreys would report the posterior probabilities of H_0, Prob [H_0 | $x_1, x_2, \ldots x_n$] 0.28 or a probability of 0.11 in the second case, based on assigning equal prior probabilities of ½ and using a conventional Cauchy prior with parameters (0. σ) on the alternative.

Neyman, with a specified type I error probability α = 0.05, would report the figure α = 0.05 in either case along with a Type II error probability. The disagreement occurs primarily when testing a 'precise' or a point hypothesis. When testing a one-sided hypothesis like H_0: $\mu \leq \mu_0$, figures reported by Fisher and Jeffreys would be mostly similar.

Jeffreys agreed with Neyman that one needed an alternative hypothesis to engage in testing. He defines a Bayes factor (or likelihood ratio) B (x) = f (x | μ_0) / f (x | μ_1) and rejects (accepts) H_0 as B (x) $\leq$ 1 or $<$ 1. He also advocates reporting the objective posterior error probabilities Prob [H_0 | x] = B (x) / [1 B 9 x)]. Berger points out criticisms of these three approaches and recommends a conditional frequentist approach.

There have been many occasions when foundations of statistical inference have been revisited by stalwarts, specially after the Bayesian methods and the supporting simulation techniques gained ground. Research workers in statistics should definitely have a look at these discussions, whenever they enter into some theoretical research area.

3.11 Value of Information

Taking a high-value or strategic decision may well justify a non-trivial piece of research to provide inputs into the decision-making process and also to evaluate the ultimate consequences or outcomes. Too often, such decisions have to be taken in the face of uncertain situations and/or consequences. In fact, decision-making exercises are carried out in several distinct situations, viz. (1) under certainty when the consequence or outcome of any possible action is exactly known and the only problem is to find out that action which results in the 'best' outcome and the search may have to be taken up in a 'large' space of all possible and, in most cases, numerous actions. (2) Under uncertainty where we have to recognize several (finitely or infinitely many) environmental conditions or 'states of nature' and each possible action has a distinct outcome for each such state of nature, giving rise to a decision matrix with rows as possible actions and columns as possible states of nature. Here we require some optimality criterion to identify the best action. And (3) under risk where the different states of nature are likely to be realized with known probabilities. The set of probabilities associated with the possible states of nature, given prior to deciding on any action, defines what may be called a prior distribution. It should be remembered that a 'states of nature' represents the set of values and/or relations that determine the outcome of any possible action.

In the foregoing discussion, we have considered decisions and actions as synonymous and the decisions are terminal. It is conceivable that a decision-maker takes the initial decision to collect some more information before taking a terminal decision. A similar differentiation is made in the context of stochastic programming (mentioned in chapter 7) between the 'here and now' approach and the 'wait and see' approach. Against this back-drop, we need to discuss 'value of information' and the associated measures of opportunity costs. To begin with, we differentiate between immediate and terminal decisions and consider both situations (2) and (3) as 'under uncertainty'. We realize that one has to pay for uncertainty and even to reduce the cost of uncertainty through an immediate decision to gather additional information before reaching the terminal decision.

Let us define the expected cost of immediate terminal action as the expected cost of the 'best' terminal action under the given prior distribution. The cost of uncertainty is the irreducible loss due to action taken under uncertainty or the expected loss associated with the best decision for the given probability distribution. The difference between the expected cost of uncertainty and cost of the best decision under certainty about the state of nature is taken as the value of perfect information. And this would depend on the particular state of nature being true or realized. Averaged over the set of possible states, we get the expected value of perfect information (EVPI).

Given a very high EVPI, a decision-maker would tend to gather information about the true state of nature. But this implies some additional cost with two components, viz. the cost of gathering information and the cost of processing the information gathered to take the terminal decision. And both these activities may proceed sequentially. Short of gathering a lot of information adequate to remove uncertainty about the true state of nature completely at a relatively large cost (which in some adverse cases may exceed the value of perfect information) one may think of gathering sample information using a proper sampling procedure to get some reasonable idea about the true state of nature. The cost of sampling will depend on the sample size and the way sample information is processed to yield the desired information about the true state of nature. The excess of the cost of uncertainty over the cost of an immediate decision to gather sample information and to use the same for reaching the terminal decision is the net gain due to sampling. Averaging over possible sample sizes, we can define the expected net gain due to sampling (ENGS). In fact, this ENGS may be examined to work out the optimum sample size.

Some experts argue in favour of using a pre-posterior analysis, making use of the sample information to calculate posterior probabilities and working out the 'best' possible terminal decision under the posterior probability distribution. This, of course, will need some assumptions and some additional computations.

For a detailed and simplified and exhaustively illustrated discussion on the subject, one may refer to Schlaifer (1959). Besides, more recent books and articles should be found quite interesting and useful to develop the research design in many contexts, even outside the ambit of conventional business research.

3.12 Grounded Theory Approach

Grounded theory, introduced by sociologists Glaser and Strauss (1967) in the context of social science research and subsequently diversified in content and fields of application, provides a completely different approach to research design. In fact, grounded theory offers almost a parallel research methodology, differing from the one discussed in this treatise and in most other writings on the subject right from the formulation of research problems and working hypotheses to the collection and analysis of data and final interpretation of results. According to Glaser and Strauss, grounded theory would provide a method allowing researchers to move from data to theory in a back-and-forth manner to come up with new theories which would be 'grounded' in data and, hence, would be specific to the context in which data were being collected and analysed. Thus, researchers would try to generate new, contextualized theories rather than relying on analytical constructs or variables from pre-existing theories. The focus was not to support or refute existing theories as explanations of observed phenomena or even to develop modifications to improve such explanations in terms of collection of data and their analyses as are desired for such purposes. With some apparent similarity with content analysis, grounded theory differs significantly from content analysis in its objectives, basic principles, data collection procedures and interpretation of findings.

Progressive identification of categories of meaning from data and their integration to build bricks of a context-specific theory to explain the underlying phenomenon in a sequential manner is the cornerstone of grounded theory. As a method, grounded theory provides us with procedures to identify categories, to make links between categories and assess distances between categories in the process, to formally recognize categories in terms of codes (which are not pre-assigned before the data appear) and to establish relationships among categories, e.g. structural conditions, consequences, norms, deviations from norms, processes, patterns and systems. It also tells us how much data we should collect and how we can combine categories. And in this context, the term 'category' deserves a special attention. As a theory, grounded theory is expected to provide us with a framework to comprehend and explain a complex phenomenon based on the data and not influenced by any pre-existing theory or any pre-conceived notion.

Fundamental tasks in the grounded theory method include identifying categories, coding, constant comparative analysis (ensuring simultaneous collection and analysis of data), theoretical sampling, theoretical saturation, memo-writing and preparing the research report. Negative case analysis, examining theoretical sensitivity and some related steps are also involved. While a research question might have been the motivation behind research, concrete formulation of the research problem continues through the entire research process. Obviously, collection of data –mostly qualitative – from diverse sources and their interim analysis before collection of more data constitute the backbone of grounded theory.

Memo-writing is a step somewhat unheard of in the usual research processes, was possibly suggested by the initial applications in social science researches and has a distinct meaning and purpose in grounded theory.

Grounded theory bears some resemblance to phenomenological study in the sense that both make use of human experiences, seek to comprehend a situation from a framework that evolves with such experiences being available and allow flexible data collection procedures. However, there exist many important differences. Empirical phenomenological researches are concerned with the world-as-experienced by the participants of the inquiry to common meanings underlying variation of a given phenomenon. Per contra, grounded theory tries to unravel "what is going on?" in terms of the core and subsidiary processes operating in the given situation. Phenomenology dictates sources of data rather narrowly only by way of verbal statements, written reports and artistic expressions of the phenomenon under study as provided by the informants who have lived the reality being investigated.

Recent times are marked by non-traditional researches into phenomena not regarded hitherto as worth any investigations and these relate to situations where data collection is a problem by itself, where no theoretical explanation of the underlying phenomenon is available, and where a reasonable understanding of the phenomenon is of great consequence to a large audience and is still awaited in the given context. Herein comes the utility of the grounded theory approach. Limitations are bound to exist and the approach has not yet found wide applications. Critics of grounded theory have focused on its supposed ability to throw up what may be justifiably called 'theory', its grounding or over-emphasis on data and its use of induction to develop context-specific knowledge about intricate phenomena. In grounded theory, researchers engage in excessive conceptualization which cannot but be subjective to some extent and defend this excess for 'sensitivity to context'.

Data analysis in grounded theory begins with coding, defined by Charmaz (2006) as "naming segments of data with a label that simultaneously categorises, summarises and accounts for each piece of data". In fact, coding procedures are theoretical schemes or conceptual templates which allow researchers to think about possible theoretical categories and their inter-relationships. It may be noted that data in this context may arise from different sources like case studies, interview reports, events and experiences described verbally and recorded subsequently, relevant literature, and the like. Thus grounded theory is compatible with a wide range of data collection techniques covering semi-structured interviews, participant observations, focus group discussions, diaries and chronicles of events and experiences, lectures, seminars, newspaper articles, internet mail lists etc. A related technique consists of conducting self-interviews and treating the material emerging as any other data.

Categories designate the grouping together of instances (events, processes, occurrences) that share central features or characteristics with one another. Categories can be at a low level of abstraction, in which case they function as concepts or descriptive labels. Thus anxiety, hatred, anger, fear and pity can be grouped together under the caption 'emotions'. Categories which are analytic and not merely descriptive are a higher level, interpreting instances of the phenomenon under study. Analytic and descriptive categories are based on the identification of relations of similarity or difference. Unlike in content analysis, categories evolve throughout the research process and can overlap.

In a study of corruption in public procurement activities and its impact on quality of performance (Sharma, 2018), several reported cases provided the initial data. The cases were coded line by line in the manuscript to capture the key points of the data, i.e. by identifying and naming each phenomenon (incident, idea or event) which captured its essence that in turn generated concepts. [Line-by-line analysis ensures that our analysis is really grounded and higher-level categories and later on theoretical formulations actually emerge from the data, rather than being imposed upon it. If we code large chunks of data, say pages in a report, our attention may be attracted to one particular striking occurrence and less obvious but perhaps equally important instances of categories can evade our notice.] Thus emerged several concepts like copied specifications, tailor-made specifications, specification rigging, framing specifications in a biased manner, lock-out specifications, etc. Coding is likely to be iterative and code titles and descriptions were changed and refined as researchers became more familiar with data and inter-relationships emerged. To enhance credibility of coding, several measures are usually contemplated: engaging a second coder and assessing the agreement between the two, referring to relevant literature to remove potential ambiguity and developing process maps to facilitate understanding of functions being analysed. To illustrate, the author (Sharma, 2018) noted that some cases described the incidence of a favoured bidder being provided with confidential information needed to make the bid meet requirements and be accepted and found that such a case had been identified by OECD as 'unbalanced bidding'.

Theoretical sampling implies collection of further data in the light of categories that have surfaced from earlier stages of data analysis. We check emerging theory against reality by sampling incidents that may challenge or elaborate its developing claims. Theoretical sampling leads to refinement of categories, whenever needed, and ultimately to saturation of existing and increasingly analytic categories. It must be borne in mind that grounded theory is always provisional and data collection and theory generation cannot concurrently proceed ad infinitum.

Memo-writing or simply memo-ing is an important part of grounded theory methodology as providing information about the research process itself and subsequently about the findings or outcomes. In the words of Glaser (1998), "memos are the theorizing write-up of ideas about substantive codes and their theoretically coded relationships as they emerge during coding, collecting and analyzing data, and (even) during memoing". In memos, researchers develop ideas about naming the concepts and relating them to one another and present relationships between concepts in two-by-two tables, in diagrams or figures or whatever makes the ideas flow and begets ability to compare. Memos integrate contents of earlier memos or ideas: they can be – at any stage – original as we write these. Memos complete with dates and headings reveal changes of direction in the analytic process and emerging perspectives, reflecting incidentally on the adequacy of the research questions. Memos make ideas more realistic and thoughts more communicable.

By the way, the initial research question in grounded theory should be open-ended, not compatible with a simple yes/no answer, should identify the phenomenon of interest but make no assumptions about it, and should not use constructs derived from existing theories.

In preparing the report of a research using the grounded theory methodology, the researcher must indicate the research questions which possibly had not been addressed earlier in regard to a phenomenon studied by some previous workers or some phenomenon identified by the present researcher. It must be remembered that grounded theory does not fall in line with the positivist epistemology where a hypothesis is to be tested as part of an exercise to comprehend an observed or observable phenomenon. It follows the interpretive line where the focus is on the data from which is to emerge a context-specific theoretical explanation of the phenomenon. While reporting on the research process, a literature review may take place only at the stage when findings have emerged to find out how and to what extent the present findings differ from and even challenge any existing theory bearing on the phenomenon.

Results arrived at through the grounded theory approach do not admit of traditional methods of validation. Validity in this context has to be judged in terms of

Fit: Context-specific data-derived theory emerging from grounded theory applications should be in terms of concepts that are closely connected with the incidents being represented. This fitness implies thorough and constant comparison of incidents to components.

Relevance: The research exercise should address the genuine concerns of the participants (who or which really provide sources of data), evoke 'grab' and not be just of academic interest. In fact, being context-specific, grounded theory should be and has been applied not to come up with general context-free theories as in any positivist study of a repetitive phenomenon.

Workability: Theory emerging from grounded theory applied to deal with a research question should explain how the question is being solved on the basis of data.

Modifiability: A modifiable theory can be altered as and when new data are compared to data already analysed. A grounded theory approach ends up with a theory that should not be characterized as right or wrong. Of course, such a theory can attract greater fit, relevance, workability and modifiability with more data.

Charmaz (2006) provides ample illustrations of the various tasks to be performed in applying grounded theory which can help potential users of this theory. However, Morse et al. (2009) argue that different researchers may not follow the same version of the methodology to suit their respective research purposes.

3.13 Ethical Considerations

Ethical considerations have assumed great significance over the last 50 or 60 years as R&D activities were taken up by many agencies – both private and public – over and above a fast-increasing number of individual research workers in academic and professional bodies. Some in the burgeoning research community were not scrupulous and did not maintain adequate honesty or transparency or commitment to contracts and the like. This created development of standards for ethics in research – both in the research process and also in the research result. In this context, the term 'ethics' takes care of legal or regulatory requirements along with moral and ethical norms and practices.

In fact, the research process right from the research design to documentation and dissemination of research results should incorporate ethical considerations to the desired extent. In the case of biological, medical and even sociological research or in any research where humans are used as subjects or units in experiments or observations, it is mandatory to seek informed consent of the participants in the investigation, to maintain strict confidentiality of information gathered or response observed, to avoid any encroachment on privacy during experimentation or observation, and to debrief such participants about the research findings before these are brought to the public domain.

In experimental or observational research, it is essential to maintain properly all raw data and associated notes about how such data had been generated and how they had been used to generate the final or smoothed or modified data to be used in the analysis stage.

Whenever some 'material' is borrowed from some source, it must be duly acknowledged. Further, in cases where permission is need to borrow such material, it is mandatory to seek such permission from the appropriate authority.

4

Collection of Data

4.1 Introduction

In this chapter, we would like to focus on the need for data in any research, be it theoretical or empirical, primarily involving qualitative analysis using nominal or categorical data and applying simple tools of logical reasoning or quantitative analysis with its usual emphasis on statistical concepts, methods, techniques and software. This is followed by a brief discussion of the two primary data collection mechanisms, viz. sample surveys and designed experiments. Instruments to be used in these two mechanisms to collect/generate primary data have been also indicated.

Data have to be collected in any empirical research as the basic inputs. Both primary and secondary may be needed in some research, while only the former may be required in some other, while some research activity may deal only with analysis and interpretation of secondary data only. Secondary data have to be collected from their respective sources, due care being taken about their relevance, adequacy for the research objective(s), accuracy, reference to the time period of interest, operational definitions of terms and phrases used and the method of data collection and validation, etc. Sometimes similar data from different sources may have to be pooled with necessary care.

Data collection mechanisms following decisions on type, volume and quality of data as indicated in the research design play a very important role in all empirical researches. We generate numerical data when we measure or count some characteristic or feature possessed by the units or individuals selected for the study. (It should be remembered that most often – if not necessarily always – these selected units or individuals constitute a *representative* sample from the population or aggregate of units or individuals in which we are interested.) On the other hand, data by way of opinions or aptitude or attitude and also those related to purely observable attributes like language, complexion, religion, social group, occupation, etc. are 'categorical', in the sense that the responses derived from the respondents by direct observation or in terms of the responses to questions or other stimuli can be put into several categories. The number of categories should be neither too small to put all respondents in one category or a very few categories, nor too large to find very few responses in quite a few categories. In some cases, there may exist a natural number while in some others no such numbers pre-exist. Even in the case of a number pre-existing, we may group them into a smaller number. All this depends on the context.

Data directly obtained are 'primary' while those gathered from already existing sources that can serve our purpose are branded as 'secondary'. In many studies both primary and secondary data are called for. Sometimes, we gather secondary data from several different sources. Before we pool such data, we must make sure that they are 'poolable' in terms of their purposes, operational definitions of various terms and phrases, collection and verification mechanisms, and the like.

Quite apart from this thread of discussion are research problems with the mechanisms for data collection. In fact, in that sense both sample surveys and designs of experiments pose research problems which can be quite theoretical or could take into account practical problems of implementation. Section 4.5 gives a brief outline of a few such research problems in connection with designs of experiments in industry and social science research, and also with sampling designs for large-scale surveys.

4.2 Collection of Primary Data

4.2.1 Sample Surveys and Designed Experiments

Primary data have to be generated, either by conducting an experiment in a laboratory or in the field, which has to be properly designed and conducted to yield the desired data, or by planning and conducting a sample survey, which has to be properly planned. Usually sample surveys are conducted with an appropriate sampling design. The basic difference between these two data-gathering exercises is that units or individuals in a survey are observed or assessed as they are, with no intervention on the units, while these are subjected to some treatment in an experiment to provide some responses which are noted in respect of the units or individuals treated.

On the other hand, secondary data which are 'primary' data already collected by some agency – public or private – on a regular or an ad hoc basis can be obtained from relevant sources. Data on socio-economic aspects of the country or some region as well as demographic data are routinely collected by public authorities and can be accessed for research purposes. In fact, if research involves data relating to the past, we have to make use of credible secondary data. Otherwise some data may also pertain to sensitive segments of the population, e.g. those engaged in defence services or the economy dealing with manufacture and deployment of defence equipment. For research on health issues, hospitals and nursing homes can provide useful data, though with some limitations on the coverage.

4.2.2 Design of Questionnaires

Most sample surveys make use of a questionnaire or a schedule among the selected units and individuals, and the responses made by the respondents, with necessary explanations when called for, are recorded. A questionnaire is a list of questions given out in a desired sequence, while an inventory/battery is a list of statements. Questionnaires can be (a) open-ended, allowing the respondents to use their own words to answer any item (question), or (b) closed-ended, asking the respondent to tick one among the response categories indicated in the questionnaire. 'Schedule' is sometimes used in this context.

A questionnaire can be (a) disguised, if it does not directly link the items with the underlying objective(s), or (b) undisguised, exposing the objective(s) to the respondents.

- As an alternative to a closed-ended questionnaire, we have a list (inventory) of statements, used quite widely.

- Questionnaires and inventories have to be carefully developed, pre-tested and debugged before use on the planned scale.
- Only relevant questions/statements which are unambiguous, non-burdensome and easy to complete should be put in a desired sequence.
- Opening questions/statements should be interesting enough to 'engage' the respondents in the exercise. Classificatory questions should usually be asked at the end. Assurance should be given about confidentiality or anonymity.
- Screener questionnaires are filled up by interviewers by observing or listening to the 'selected' respondents.

Unless imperative, questions relating to sensitive or confidential information should be avoided in terms of direct questions. In such situations, randomised response techniques may be taken advantage of to derive estimates of population parameters. (This technique does not help in getting 'good' responses from individual respondents.)

The quality of data collected through the use of a questionnaire is affected by the way the questionnaire is canvassed or administered, to individuals or groups, with or without interactions among the respondents, seeking instant answers or considered responses with adequate time to ponder over the questions. Of course, biases and errors on the part of the respondents will seriously affect the quality. While biases are usually intentional and uni-directional (resulting consistently in under- or over-statements), errors are usually unintentional and can work in both directions. Such biases and errors are sometimes induced by the lack of clear and correct interpretation of each item in the questionnaire or schedule.

Concepts like income or expenditure during a month on food or literacy or employment/unemployment status or economic establishment or even the highest level of education for an individual may admit of more than response – none absolutely wrong – and we have to specify the concept in terms of a standard or an accepted operational definition. Otherwise, genuine variations among individuals may be masked and/or artificial variations may show up. An innocent example is when 'age (in years)' is asked as a common information item in respect of the respondent. The latter can take 'age' as 'age last birthday' or as 'age next/nearest birthday'. The consequence is that for two individuals born on the same date the recorded ages could differ by two years, and for two other individuals, one younger than the other by one day less than two years, recorded ages may be the same.

And we all appreciate that 'variation' is the basic entity that we want to describe, quantify, compare and explain.

4.2.3 Scaling of Responses

Responses are often scaled as scores to facilitate aggregation across items and over respondents and to make responses in different situations comparable. We have the Likert scale, the rank-order scale, the semantic differential scale, etc. Likert scaling with usually an odd number of response categories is most often used.

Let us briefly discuss Likert scaling, which is widely used in many enquiries where we deal with categorical responses to certain questions. Attributes like perception, attitude, honesty, sincerity are *unobserved latent variables*. The response of an individual to a statement or a question relating to such an attribute is generally recorded as belonging to one of several pre-fixed categories. To a question on how one felt about some programme on research methodology, a respondent can state *Fair* or *Good* or *Very Good* or

Excellent. (One may add a fifth category, viz. *nothing particular* or *undecided.*). Given a statement like *My colleagues/peers give me full co-operation in discharging my duties* in an organizational climate survey, possible reactions could be *completely disagree, disagree, cannot comment, agree* and *completely agree.*

To proceed further to analyse the *ordinal* data, we need to assign scale values to each category, assuming a certain probability distribution of the underlying latent variable over a support (a, b). Usually we have an odd number (5 or 7) of categories and we assume a normal distribution for the underlying trait or latent variable. We can have any number of categories and can assume any other trait distribution. The upper class boundaries for the latent variable are, say, $x_1, x_2, \ldots x_k = b$ (maybe ∞). The task is to estimate these unknown boundaries in terms of the observed frequencies of responses in the corresponding categories and then find mid-points of the classes.

Given frequencies $f_1, f_2, \ldots f_k$ in the k categories with cumulative frequencies $F_1, F_2, \ldots F_k$ respectively, we equate F_i to $\Phi(x_i)$ to get x_i from a table of Φ values (left tail areas under the assumed Normal trait distribution with mean 0 and SD 1). It can be shown that x values thus determined are reasonably good estimates of the unknown boundary points for the pre-fixed number of categories and, hence, of intervals for the trait or latent variable. Intervals for the trait are replaced by means of the trait distribution truncated between the boundaries for the intervals.

Assuming normality, the truncated mean for the class (x_{i-1}, x_i) is given by

$$s_i = (\varphi_{i-1} - \varphi_i) / (\Phi_i - \Phi_{i-1})$$

The first class for the Normal trait distribution is $(-\infty, x_1)$ while the last class is (x_{k-1}, ∞) with

$$\Phi(-\infty) = 0, \Phi(\infty) = 1, \varphi(-\infty) = 0 = \varphi(\infty).$$

Sometimes, scale values are taken as equidistant integers like 1, 2, 3, 4, 5 or 2, 4, 6, 8. These then become data- and model-invariant. We must note that trait intervals corresponding to different response categories are not generally equal. In Likert's scaling, the scale values are data-dependent and will depend on the particular set of observed frequencies as also on the trait distribution assumed.

An Example

Quality of service offered at a newly opened customer service centre judged by 60 visitors

Grade	Frequency	Cumulative frequency
Poor	3	3
Fair	14	17
Good	26	43
Very good	13	56
Excellent	4	60

Assuming an N(0,1) distribution for the underlying trait, viz. *Perceived Quality,* upper class boundaries become −1.65, −0.58, 0.55, 1.48 and ∞. Thus, scale values are −2.06, −1.04, 0.01, 1.02 and 1.90. If we take an arbitrary mean of 3 and an SD of 1, these values become 0.94, 1.96, 3.01, 4.02 and 3.90 and are not equidistant.

In a situation where the trait can vary over the range 0 to 5, and we can justifiably assume a uniform distribution of the trait, the scale values will be 0.25, 0.83, 2.50, 3.45 and 4.67 respectively.

4.2.4 Survey Data Quality

Quality of data collected through a survey implies features like relevance, accuracy, adequacy and (internal) consistency.

These features are affected by the design of the questionnaire and also reliability and validity of the instrument (the questionnaire or schedule) used to collect the data. Reliability and validity are measurable. These are measurable features though the measures are neither unique nor applicable to all situations.

Validity relates to the question: does the instrument measure what it purports to measure?

Reliability examines the consistency or stability with which the instrument measures what it does measure.

There exist several facets of validity and reliability. Thus, face validity, concurrent validity, predictive validity and construct validity are spoken of in the context of validity. Similarly, we have test-retest reliability or split-half reliability. Face validity is ensured by including only pertinent questions which can be answered by the respondents, given their level of understanding and ability to communicate their feelings or views or knowledge. Concurrent validity has to be established in terms of agreement between responses obtained by the present questionnaire with those obtained from a similar group (if not the same group of respondents) canvassed for the same purpose. It may be difficult to go by this requirement. And predictive validity calls for data on some characteristics of the respondents like performance or action on some future occasion with which data from the present survey will be correlated. Test-retest reliability examines the agreement by way of correlation between responses given in two consecutive administrations of the test separated by a reasonable time gap. Unfortunately, this possibility is most often ruled out in large-scale surveys. Split-half reliability is measured by the correlation between responses (converted to scores whenever needed) in two halves of the questionnaire matched in terms of difficulty levels and other relevant considerations. This implies some questions which may seem to be duplicates of one another.

In content analysis, useful in social science research, data are deciphered by scholars or coders from observations (including incisive ones) and even perusal of writings/artefacts/images etc., as these unfold over time or across authors. Such data are usually unstructured and have to be made amenable to analysis.

4.3 Planning of Sample Surveys

4.3.1 Some General Remarks

Censuses and sample surveys have been established instruments for collecting data bearing on different facets of a country or region, its people and society, its economic and political status and similar other aspects for extracting information pertaining to the level prevailing in a period of time and also the changes that have crept in over time. India took a lead under the guidance of Professor P.C. Mahalanobis in using sample surveys to derive credible information relating to diverse entities, starting from area and yield of jute, through socio-economic data, public preferences to migration and tourism, to a wide array of entities which have become quite important for planning and evaluating various plans and programmes. India also made pioneering studies on the planning of surveys, taking care of theoretical aspects concerning the use of stratification, of ancillary information, of working out estimates of sampling errors through inter-penetrating networks of subsamples and the like as well as of practical problems related to cost.

Censuses and (sample) surveys of some kind or the other have been carried out since the good old days to meet various administrative or political needs. However, the first scientific large-scale survey was planned and executed by the famous mathematician Pierre Simon Laplace (who, of course, made a significant contribution to statistical induction), who in 1802 tried to estimate the population of France on the lines of John Graunt of England and on the basis of the annual total number of births, assuming a stable CBR (crude birth rate), using what may now be called two-stage sampling design. He even worked out a credible interval for the estimated population using inverse probability.

Surveys, obviously, are multi-item in the sense that more than one item of information is collected from a responding or participating individual. Such information items may be simply collected through a questionnaire or through tests and examinations conducted on the individuals. In fact, the quality of survey data in the latter cases will depend a lot on the survey instruments used and the skill of the interviewer or investigator.

The theory of sample surveys was initially focused on a single variable, or better, a single parameter (summarizing the distribution of this variable) which was to be estimated from the sample data with a reasonable degree of accuracy.

Important concerns in the planning and conduct of a sample survey include, among others, delineation of the population of interest, identification of the parameters to be estimated, determining the total sample size to enable some stipulated level of accuracy, stratification of the population based on available information to be properly used, allocation of the total sample size over the different strata, choice of the appropriate estimator, etc. Leaving aside these problems of sampling design, data collection by recording measurements or noting responses causes serious problems about data quality.

Given that most – if not all – sample surveys are meant to provide estimates of several parameters that relate to several variables on which information is collected, developing a proper sampling design becomes quite problematic, particularly when we have little information available on the likely distributions of the different variables on which we get responses.

There have been many abuses of sample surveys that have serious impacts on common people. In fact, quite a few so-called international organizations – not necessarily official – come up with findings based on a completely incorrect design of sample survey and without any care for the quality of responses and put such findings on their websites. Not all viewers of such sites are expected to be discerning and may be falsely led to believe such findings, which – in reality – constitute a disdainful dis-service to sample surveys.

4.3.2 Problems in Planning a Large-Scale Sample Survey

Planning a large-scale sample survey involves the following, among other issues.

Problems in Developing a Sampling Frame

Identifying the units constituting the population of interest and, hence, getting the size of the population (as the number of ultimate units yielding responses)

Determining the sample size to estimate the population parameters with some desired level of precision, separately for different segments of the population, whenever needed

Developing a suitable sampling design that can yield estimates of the parameters of interest with the desired level of precision and for different specified segments of the population (usually different geographical regions)

Problems in Use of Stratification

It may be mentioned that most often a multi-stage stratified sampling design is adopted, and this implies:

1. identification of different stages in sampling
2. choice of suitable stratification variable(s) and
3. allocation of the overall sample among the different strata in a given stage.

Of course, determining the total sample size has to be an important exercise.

In these choices, one complicating factor is the presence of more than one parameter to be estimated, with prioritization not always admissible.

A multi-stage stratified sampling design (sometimes with stratification in two stages) is often adopted in large-scale sample surveys. And this is the usual practice of the National Sample Survey Office (India). Further, this design has more than justification in situations where, within a certain stage, units are quite heterogeneous and one should use some homogeneous grouping to achieve representativeness.

The major point of discussion in stratification concerns the choice of the stratification variable(s) and the partition points to separate the different strata. In fact, the number of strata to be recognized is itself a matter of decision. It may be added here that if natural boundaries exist within the population to be covered, like districts or sub-divisions or blocks or some similar aggregate, we tend to make use of them as strata. However, such natural boundaries are not appropriate for the purpose of ensuring within-stratum homogeneity in most situations and for most of the survey objectives.

A real-life example may be worth reporting. For the Annual Employment-Unemployment Surveys conducted by the Labour Bureau, Ministry of Labour and Employment, Government of India, households within a selected village were stratified on the basis of household consumption expenditure level (information being collected during the house-listing operation) with quintiles corresponding to (second-stage) strata. The present author, in his capacity as Chairman of the Expert Group for the surveys, criticized this stratification scheme on the grounds that in a household with only one member in the labour force employed in a high-salary regular job and the remaining members seeking but not getting employment, the per capita consumption level may be high, whereas in a second family where all eligible members are employed but in low-paid jobs this consumption level may be poor. In fact, the employment situation among household members and consumption level may not have any necessary linkage. On the advice of this author, the strata were subsequently formed on the basis of the number of members in the eligible group (aged 15 and above and engaged in or seeking a job). Here again, the strata were formed somewhat arbitrarily as 0, 1–2, 3–5, 6 and above.

We must bear in mind that arbitrariness can hardly be avoided in real life.

Sometimes, two or more stratification variables have been simultaneously considered for the purpose of constructing strata. For example, we can think of a classification of districts or even blocks on the basis of, say, female literacy rate and percentage of people belonging to Scheduled castes and tribes. To determine cut-off points, one can think of the average rate and average percentage for a bigger ensemble like a district if blocks are to selected or a state if districts are to be selected, and proceed with a 2×2 classification to yield four strata, by considering less than or equal to average and more than average corresponding to the two classes for each characteristic (stratification variable). It is just possible to think of more strata in terms of a 3×3 or 4×4 classification, taking into consideration the three quartiles for each characteristic. Fortunately, distributions of each such characteristic are available at least for the census years. If a cell shows a very low frequency, it can be merged with a suitable neighbouring cell.

Sample Size Determination

The sample size required to estimate an unknown population parameter can be obtained from a specified relative standard error of the estimate or from a specified length of the confidence interval (for the population parameter) with a specified confidence coefficient. Large sample normality is a common assumption in the latter case, though no such assumption is involved in the first. The two procedures are, in essence, the same. The length of the confidence interval with a confidence coefficient of 95% approximately is precisely four times the RSE. This length can be specified to yield the desired sample size.

Thus to estimate a proportion whose prior guess is, say, 0.2, the confidence interval for RSE = 0.10 (10%) with a 95% confidence coefficient is (0.16, 0.24), while this shrinks to (0.18, 0.22) for an RSE = 0.05 (5%). However, halving the RSE means increasing the sample size four times.

Eventually, sample size must be fixed on a consideration of precision (measured in terms of RSE or alternatively the length of the confidence interval with a specified confidence coefficient) and feasibility (in terms of available resources and of limitations of population size).

Further, we must remember that the sample size, meaning the number of responses (responding units), used in estimating a population parameter of interest may not be the same as, say, the number of households or establishments where the units are observed. The latter number may be larger than or even smaller than the former or the sample size. And the relation between these two numbers cannot be pre-judged, making it difficult to fix the latter number, which is important for drawing up the sampling design.

Cluster sampling is commonly use in large-scale surveys and selecting an additional member from the same cluster adds less new information than would a completely independent selection. The loss of effectiveness through the use of cluster sampling, instead of simple random sampling (for which only usual formulae for RSE are valid) is the 'design effect'. The design effect is basically the ratio between the actual variance, under the sampling method adopted, and the variance computed under the assumption of simple random sampling.

The main components of the design effect are the intra-class correlation and the cluster sample sizes and the effect is then DEFF = 1 + α (n − 1) where α is the intra-class correlation for the statistic in question and n is the average size of a cluster.

The DEFT (square root of DEFF) shows how much the sample size and consequently the confidence increases. The DEFT is to be used as a multiplier. It usually varies between 1 and 3 in well-designed surveys. It is quite difficult to decide on this multiplier in any national survey, the consequence of this multiplier on the sample size and hence the cost of the survey, being a primary consideration to be taken into account, dealing with several statistics and with little information about their intra-class correlations. Whenever incorporated in any national survey, it has been chosen only arbitrarily, at best referring to its value used in some country for a more or less similar survey.

In direct response surveys, the sample size required to estimate a population parameter with a specified error in estimation can be found from the condition Prob. [| t − Y | ≤ f Y] ≥ 1 − α where t is an unbiased estimator of the parameter Y, α is a fraction usually quite small and α is a pre-assigned small quantity., which gives n = (N− 1) / N CV² / α f² for SRSWR and = N / [1 + N α f² (100 / CV²)] for SRSWOR where CV is the coefficient of variation of the estimator t (as given by Chaudhuri in 2010, 2014 and Chaudhuri and Dutta in 2018). The expression for sample size for general sampling strategies has also been indicated by Chaudhuri and Dutta (2018). A solution to the sample size problem in the case of a randomized response survey is yet to be found.

In both the approaches, which are fundamentally the same, we need to use the CV of the estimator either from past records available before the survey is undertaken or obtained from a sample of some arbitrary size through resampling methods and then used to determine the sample size for the actual survey being planned.

4.3.3 Abuse of Sampling

The Charities Aid Foundation compiles the World Giving Index (WGI) using data gathered by Gallup and ranks 153 countries according to how charitable their populations are. The Foundation claims that the WGI is the largest study ever undertaken into charitable behaviour around the globe. The survey found that around the globe happiness was seen as a greater influence on giving money than wealth.

What disturbs the author is the starkly inadequate sample size and the improper sampling design adopted for the study. In most countries surveyed, only 1000 questionnaires were canvassed to cover a 'representative' sample of individuals living in urban centres. In some large countries such as China and Russia samples of at least 2000 were covered, while in a small number of countries, where polling is difficult, the poll covered 500 to 1000 people, but still featured a 'representative' sample. Respondents were over 15 years of age and samples were 'probability-based'. Surveys were carried out by telephone or face-to-face conversation, depending on the countries' telephone coverage.

Gallup asked people which of the following three charitable acts they had undertaken in the past month:

- donated money to an organization
- volunteered time to an organization
- helped a stranger, or someone they did not know, who needed help.

Giving money or time to an organization could include political parties or organizations as well as registered charitable societies, community organizations and places of worship.

The sample size used is far from any norm to provide even a crude estimate for the country as a whole. Representativeness claimed is just incredible, and cannot be judged in the absence of any information about the method of selection and the 'frame' used for the purpose of selection. Restricting the sample to an age-segment and a residence location will not enable the 'quota' sample to furnish an estimate for the population as a whole. And, of course, questions arise about the 'money- or time-giving activity' itself, since giving money or time to a political party or organization may be an activity prompted by fear or force and may not reflect an individual's 'charitable behaviour'.

Websites like this one mislead common people by the country ranks they provide and the interpretations made without discretion or without a scrutiny of the methodology used.

Such websites are not uncommon.

4.3.4 Panel Surveys

In some research problems as also in ongoing surveys to generate data for socio-economic analysis, we need to follow up a specified group (sometimes referred to as a cohort or a panel) of units over time to track changes in some parameters of interest among these units. Thus, an Annual Survey of Manufacturing Industries is carried out in which a given set of units is visited every year to collect data on several parameters including various inputs and outputs and eventually yielding the gross value addition by each unit. Changes over the years are indirectly indicative of changes in national income (gross domestic product).

Recently, a Survey on Monitoring of Outcomes (of programmes meant to benefit potential mothers and children) on Children (SMOC) was being planned by UNICEF to track different growth trajectories of children and to relate differences in levels over age. The following design choices were available, and once a choice was made, appropriate methods of estimation of both (1) current levels of some indicators to trace changes over time and (2) change in any chosen indicator since the last time it was noted were to be considered. The first is a cohort or longitudinal survey which selects a single sample from the population of interest and studies the same units (households and beneficiaries) over consecutive rounds. Chances of attrition due to migration would be high, quality of revisit data could be biased and incoming units would continue to remain outside the investigation. The second is a rotational panel survey, which produces reliable estimates of change as well as cross-sectional level estimates, Here, a designated part of the sample is replaced by new units at a regular, pre-specified period. The third is a set of repeated cross-sectional surveys which selects a new sample in each round to track the movement of population characteristics over a long period of time.

The second provides a good compromise between longitudinal or cohort surveys and repeated independent cross-sectional surveys. It is expected that each rotation panel produces an unbiased estimate, so that the average of the panel estimates provides the desired estimate of the level. To take care of rotation bias, an AK composite estimate of level can be built up as follows

$$\text{Est total } Y = \tfrac{1}{4}\,(1 - K + A)\,Y$$

where A and K are so determined that the efficiency of the estimate is maximized and D is the estimate of the change based on the common rotation panels between survey round c and c − 1.

Panel surveys provide longitudinal data required in tracking or follow-up studies to delineate trends and cyclical fluctuations in some variable of interest. These are conducted in different rounds or waves covering the same sampling units in each wave, except when a new group is designed to replace an existing group within the set of sampling units existing in the previous wave. This is the case for a rotational panel.

Panel surveys could be broadly of three types. Firstly, we can consider a cohort of individuals like children admitted to the same starting class in a primary school (or even in the totality of schools within a school district) and follow them up year after year in respect of their progress in education till none of them continues in school education. Such a cohort study will also reveal the number of children who drop out of school, who stagnate in the same class/grade and who leave the school(s) in the study to join some other school(s). The consequence is a decreasing number of units or children in successive years. Somewhat permanently staying in the study would be households and we can have a panel of households followed up during successive periods – months or quarters or years – in respect of their per capita consumption of, say, food items or recreation services. In fact, many such panel surveys with a fixed panel of households are being used in many countries. An annual survey of registered manufacturing industries is also conducted in India and certain other countries. While the panel of industries is fixed, the panel may change not by design but by default whenever some existing industry is closed down or some new industries come up. As distinct from this second type of panel survey are rotational panel surveys where a specified fraction of the initial panel is replaced by a new sub-panel during each wave. This rotation avoids some biases in fixed panel surveys and provides a better representation of the population over the waves taken together. An important element in the design is the fraction to be replaced or, alternatively, the proportion matched between successive waves

Panel data are not synonymous with time-series data. Panels correspond to sampling units, usually at the micro-level where the units may change slightly in size or some other dimension. A time series may be in terms of data at the macro-level with even significant changes in such parameters. Thus, a time series on GDP or total employment in services or total export earnings illustrate time series but not panel data.

4.4 Use of Designed Experiments

4.4.1 Types and Objectives of Experiments

Experiments are carried out to generate empirical evidence (not based on theory) and to provide pertinent information relating to some phenomenon or phenomena (including systems and operations) under investigation or bearing on some objective(s) to be achieved. Experiments range from simple and quick observations, through design development and administration of a test or a treatment, to complicated and even hazardous ventures into unknown tracts and entities. Simple laboratory experiments to identify and estimate properties of unknown substances, bioassays to determine the relative potency of a new biological preparation compared to a standard one, agricultural field trials with different manures or varieties or irrigation practices to find out their relative merits and deficiencies, clinical trials with different drugs or treatment protocols and costly experiments in space research etc. illustrate the panorama of experiments.

While a wide array of experiments have been carried out over time, planning experiments to yield results that would provide a basis for inductive inferences started in the context of agriculture in the 20th century. Since then planning an experiment and analysing experimental results have become the key elements in any scientific and technological investigation today.

It is useful to distinguish between two types of experiments, viz. absolute and comparative. An example of the first type could be an experiment to determine the electric charges of an electron or to find the mean root-length of a plant species treated with a specified dose of an antibiotic compound or an experiment to find out if a given dose of a certain biological preparation does produce muscular contraction when applied to a mouse or an exercise to find out how an individual reacts to a given artificially created situation. It is for such experiments that the theory of errors was originally devised. Repeated outcomes of such an experiment do not agree exactly with one another and the problems considered were to obtain best estimates and associated measures of reliability from sets of observed outcomes.

On the other hand, in a comparative experiment two or more experimental conditions are compared with respect to their effects on a chosen characteristic of the population under study. Suppose a metallurgical engineer is interested in finding the effects of two different hardening processes, viz. oil quenching or brine quenching, on some metallic alloys (steel). The engineer subjects a number of alloy specimens to each quenching medium and measures the hardness of each specimen after quenching by a specified method.

The primary objective of an experiment could be some or all of the following:

(1) Identify factors or independent variables (out of the whole set of postulated or reported factors) which can be controlled by the experimenter and which have significant effect on the response(s) or dependent variable(s) and should be screened for subsequent deletion or retention in the course of future experiments. In this way, we get what are usually known as screening experiments.

(2) Estimate (from the data arising out of the experiment) a suitable model (relation or causal connection between the response variable(s) and a set of factors or predictor variables. To be specific, it could be a relation connecting a dependent variable like time to rupture of a metallic structure (which is just recorded) and an independent variable like stress applied to the structure (which is controlled by the experimenter).

(3) Test a hypothesis which could involve a comparison between two or more mean values or proportions or variances. These parameters to be compared could relate to the same response variable recorded for different levels of the same factor or for different level combinations of two or more factors. These are comparative experiments and are the most commonly used experiments.

(4) Determine the level or level combination of the factor(s) based on the estimated model usually represented by a response curve or surface – mentioned in objective 2 – that yields the 'best' value of the response variable. Thus, we may have an experiment that seeks to find the alloy that has the maximum time to rupture for a certain high stress. These are optimum-seeking or optimization experiments. The word 'best' may correspond to either an unconstrained maximum or minimum of a single response variable (which itself could be a composite of several response variables) or the maximum or minimum of one response variable, subject to some constraint(s) of some other response variable(s), e.g. maximize % yield subject to some specified minimum level of quality in terms of purity, etc.

Thus screening the factors likely to influence the response(s), estimating a factor-response relation, comparing some effects or other parameters for their possible significance and seeking the optimum factor-level combination are the primary objectives of an experiment, which should be designed accordingly.

The design for an experiment is basically a set of decisions, depending on the objective(s) of the experiment, about

Choice of factors or independent variables

Choice of levels of each factor – both the number and the values or positions inside the feasible range

Choice of the response variable(s) which should be amenable to measurement and analysis and can lead to the achievement of the objective(s) of the experiment. If the response variable is not directly measurable, we have to choose an appropriate substitute variable with a known relation to the response variable to be studied.

The design should also take account of the precision of the estimates or confidence level associated with any inference to be made. All experiments involve costs linked to materials, personnel, equipment, time etc. and the total cost cannot exceed a certain prescribed amount. Accordingly, the design will have to state the number of experimental units or runs to be considered. Also, the allocation of experimental units across the levels or level combinations should be specified in the design.

As indicated in the previous chapter, the three basic principles of an experimental design are (a) randomization, (b) replication and (c) local control, and these were initially meant for agricultural and some biological experiments. Randomization, implying random allocation of treatments to experimental units and random order in which individual runs or trials are to be conducted, is necessary to validate the application of statistical tolls for analysing the results of an experiment, ensuring that observations and associated errors are independent – an assumption made in analysis. Replication, implying repeated trails with the same treatment on different experimental units, is required to provide an estimate of error in the result corresponding to any treatment. If a response to a particular treatment is denoted by y, then the error associated with the average response is given by s_y^2 / n, where n is the number of replications. The third principle is desirable particularly when responses from different experimental units are likely to be affected by some associated variables (not separately recognized) and a comparison of responses to different treatments gets affected by variations in such variables. Here experimental units can be grouped into relatively homogeneous blocks so that units within a block receiving different treatments can generate valid comparisons over treatment effects.

We recognize the fact that age may influence the response and so we decide to apply medicine A to two adults belonging to the first age-group selected at random, B to two others, again selected at random, and medicine C to the rest. A similar principle is adopted for block 2 and in block 3 we obviously apply medicine C randomly to the three individuals in the group. In this case we will partition the total variation among the 15 responses into three categories, viz. between blocks, between medicines and error (unexplained residual).This design is referred to as a randomized block design.

In the above three designs, we were really interested in analysing random variations in responses attributable to a single factor controlled at several levels, with some part of the assignable variation explained by blocking or grouping of experimental units. However, in most experiments we allow several factors affecting the response to be controlled at several levels each.

4.5 Collection of Secondary Data

There has been a deluge of data in recent times on diverse phenomena generated at different sources through different mechanisms and stored in different databases using different technologies on different platforms. We come across data which are numerical or categorical, structured or otherwise, have different degrees of veracity and carry different weights in an exercise to combine them for the study of a group of related phenomena or even a single phenomenon of interest. These are all primary data collected by different agencies – some public, others private, exclusively for the purpose under investigation or as a by-product in some investigation on a related issue. And the research worker can have access to all these databases quite conveniently.

In many empirical researches, data pertaining to various aspects of the economy, society, the environment and state functioning or the like are to be compiled from various concerned sources. And for some data required by the research worker, there could be more than one source – in both public and private sectors. Sometimes some data are available from one source while data related to some other aspect of the phenomenon to be studied can be obtained only from a different source. And the two data sets may have used different operational definitions of the same terms and phrases, may have used different data collection methods, may have deployed investigators and supervisors who are equally trained or experienced and so on. All this makes two data sets not strictly 'poolable' to form the eventual database for the research at hand. In most such cases, simple adjustments to reconcile the two data sets may not suffice.

4.6 Data for Bio-Medical Research

Since most bio-medical researches are evidence-based and the level of evidence differs from one source to another or from one mode of data collection to another, it is important to distinguish between the following modes. Of course, the choice of the appropriate mode of experimentation or data collection depends on the objective(s) of the research.

Case-control studies – to estimate the odds of developing the disease / pathological condition under study. Also to detect if there is any association between the condition and the associated or suspected risk factors. This method, which is less reliable than cohort studies or randomized controlled studies, cannot estimate the absolute risk (i.e. incidence) of a bad outcome. The method provides a quick way of generating data on conditions or diseases with rare outcomes efficiently.

Cohort studies involve a case-defined population having a certain exposure and/or receiving a certain treatment followed over time and compared with another group not so exposed and/or treated. Such a design could be prospective (exposure factors identified in the beginning and a defined population is followed into the future) or retrospective (past records for the identified population used to identify exposure factors). This design is used to establish causation of a disease or to evaluate the outcome/impact of a treatment, when controlled randomized clinical trials cannot be conducted. The method is less reliable than randomized controlled studies since the two groups could differ in ways other than the variable under study. Further, this method requires large samples, can take too long a time and is inefficient for rare outcomes.

Randomized controlled studies with two groups, viz. the treatment group receiving the treatment under study and the control group with no treatment (placebo) or standard treatment, patients being assigned randomly to the groups: regarded as the 'gold standard' in medical research with no bias in choice of patients for treatments, so that group differences can be reasonably attributed to the treatment(s) under study, this method is widely accepted in medical research. One important limitation of this method is that it cannot be used when a placebo or an induced harmful exposure/experience is unethical.

Double blind studies – neither the physician nor the patient knows which of several possible therapies or treatments a patient is receiving, e.g. allowing the patient to pick up one of several similar pills (known only in records to a third party). This is the most rigorous design for clinical trials, since it reduces or removes the placebo effect.

Meta-analysis – combining data from several studies to arrive at pooled estimates of treatment effectiveness and to increase sample size in case-control or cohort studies. Of course, care has to be taken about the way data were collected, including definitions and measures of responses, before pooling data from different studies with the same objective. In fact, publication bias (implying studies with no or little effect are usually not published and only those with some significant effects are published, and therefore being pooled will lend bias towards the significance of effects in the pooled data) and possible variations in design in the various studies may cause problems.

Systematic reviews are comprehensive and critical surveys of literature to take note of all relevant studies with the highest level of evidence, whether published or not, assess each study, synthesize the findings and prepare an unbiased, explicit and reproducible summary to evaluate existing or new technology or practices. The result of synthesis may differ from individual studies and thus provides a more dependable guide. Such a review is different from meta-analysis, which is quantitative in character. And systematic reviews are different from reviews of the literature as one of the primary steps in any research. In the latter no attempt is usually made to synthesize the findings of previous studies, the focus being on identifying gaps in analysis or limitations of findings.

4.7 Data for Special Purposes

Data required for research into such phenomena as the impact of corruption in public-sector contracts on the quality of project performance or, say, the phenomenon of involvement of political personalities and/or bureaucrats at different levels in public distribution systems and negatively affecting the potential beneficiaries are not easy to collect, since these are hardly documented anywhere and are not obtainable by direct answers to questions raised by an investigator. Sometimes, relevant data parts may be scattered over a whole range of documents prepared by different agencies with different ulterior motives and having different degrees of credibility. We sometimes refer to investigative journalism as unearthing crucial and confidential items of information as may be required for research into the underlying phenomenon – as distinct from a high-level probe to identify and/or nab the wrongdoers involved.

Primary data may have to be collected from individuals or institutions involved in or affected by or even benefitting from some decisions or actions which are not recorded or easily accessible even when recorded. Adequate protection for confidentiality of information has to be provided to the informants who agree to part with some information. Extensions of randomized response techniques may be had recourse to for this purpose.

If some researcher wants to study the nature and extent of rights violation by state agencies among members of a specified group like inmates in prison or children in a rescue home or mentally retarded adults in a mental asylum, the researcher may require interactions with members of the group, who are not always expected to provide a complete and correct picture, and people running or managing these institutions may also provide evasive or defensive answers. And the researcher has to clearly identify features or situations or factors which can throw light on violation of rights. And the way the questions or statements should be framed and the manner in which responses as they come should be recorded deserve special attention.

4.8 Data Integration

Data integration involves combining heterogeneous data sources, sometimes called information silos, under a single query, providing the user with a unified view. Such a view is often needed in empirical research that draws upon data from disparate sources, not compatible as such. As given out in Wikipedia, such a process becomes significant in a variety of situations. Thus, two similar companies may need to merge their databases or research results from different domains or repositories may need a combination in some medical or biological research. With the phenomenal increase in volumes of data bearing on a given subject or issue or problem and the growing need for collaboration among research workers, data integration has become a bad necessity, particularly in researches linked to repetitive experiments. Integration begins with the ingestion process, and includes steps like cleaning and ETL (extract, transform, load) mapping. It ultimately enables analytics to produce effective, actionable business intelligence.

Probably the most well-known implementation of data integration is building an enterprise data warehouse. The benefit of a data warehouse enables a business to perform analyses based on the data stored in the data warehouse. This would not be possible to do on the data available only in the source system., which may not contain corresponding data, even though the data are identically named while they really refer to some different entities. This approach is also not that suitable where data sets are frequently updated, requiring the ETL process to be continuously re-executed for synchronization. Data hub or data lake approaches have been recently floated to combine unstructured or varied data into one location.

From a theoretical standpoint, data integration systems are formally defined as a triplet (G, S, M) where G is the global or mediated schema, S is the heterogeneous set of source schema and M is the mapping queries between the source and the global schema. Query processing is an important task in data integration.

Data integration involves several sub-areas like (1) data warehousing, (2) data migration, (3) enterprise application / information integration and (4) master data management.

Data integration offers several benefits to business enterprises in building up business intelligence and, therefore, in business research. As mentioned earlier, umpteen experiments are being carried out with more or less similar objectives on the same phenomenon in medicine or bio-informatics or geo-chemistry or astro-physics or some facet of climate change in different laboratories across the globe and quite often the resulting data sets are put in the public domain. To get a unified view of the data – sometimes seemingly disparate and discordant – we need to integrate the data, taking care of possible differences in operational definitions, designs of experiment, methods of measurement or of their summarization to come up with indices that are put in the data sets in the public domain. Thus, in such situations, cleaning the different data sets should have a significant role before the ETL process is taken up.

It may be borne in mind that data integration does not result in metadata. In fact, metadata means data about data, providing detailed descriptions and contexts of the data in a single data set to facilitate the search for appropriate data, to organize such data and to comprehend such data. Details like who created the data, when the data were last modified or even visited, who can access and/or update, exemplify metadata. There are three main types of metadata, viz. (1) descriptive facilitating discovery and identification, (2) structural, indicating how compound objects are put together, and (3) administrative, facilitating management like access, rights and reproduction tracking and selection criteria for digitization of a data source.

5

Sample Surveys

5.1 Introduction

The need for data for planning, monitoring and evaluation exercises has increased tremendously in recent years. Data of different types are called for at different levels of aggregation (or, rather, dis-aggregation) at regular intervals or as and when needed. All this has given rise to specialized agencies for planning and conducting national, state, district and even smaller regions where sampling designs have to be so developed that timely information on a wide array of issues can be collected and processed for a variety of purposes like the needs of administration as well as the needs of research workers. And quite often challenging problems have to be faced and solved reasonably well.

The quality of data has been a concern to all those connected with sample surveys. And several different suggestions have been made to improve data quality, including legislative measures to do away with non-response, adequate training of field staff and adequate supervision over field-work, use of palm-top computers by field investigators to record right in the field the data provided by the informant or collected from relevant documents, development and use of software to check internal inconsistencies, missing observations, outliers and similar other deficiencies, and so on. It has also been appreciated that whenever data are in terms of responses provided by the selected individual or, in his/her absence, by some informant who may not possess correct and complete information sought from the selected individual, data quality depends a lot on the respondents' ability to provide correct and complete responses, their willingness or refusal to part with the information asked for, and their appreciation of or misgivings about any likely impact of the information they provide on their interests. And in this context, an essential tool to improve data quality is an awareness campaign to convey the utility and necessity of the survey along with the absence of any adverse impact of providing truthful responses as needed for the survey.

On the theoretical front, application of some known results in the method of sampling, the method of estimation of parameters of interest and in obtaining satisfactory ideas about the accuracy or otherwise of the estimates along with the need for developing sampling designs in non-traditional problems pose interesting research problems. Some problems of allocation in using stratified sampling for multi-parameter surveys, estimation of parameters for small areas with sparse data including situations with no relevant data, designing rotational panel surveys to estimate levels as well as changes in levels of some parameters with associated problems of using composite estimators and their standard errors and the like have been considered by several researchers. However, more remains to be done.

This chapter is not designed to discuss advanced topics in survey designs and parameter estimation along with relevant results about their relative merits and shortcomings. The focus here, as is the tenor behind this book, is to highlight briefly certain methods, techniques and results which are of great relevance and importance to the execution of sample surveys as well as to draw the attention of research workers to problems that still await useful discussion and complete resolution. While the problem of developing suitable sampling designs should get due importance, problems related to implementation of a certain type of design in a particular context justifies equal attention. Just a few of such issues have been mentioned in this chapter.

Modern survey sampling, if we date it from Kiaer (1897), is now more than a century old. Early survey statisticians faced a challenging choice between randomized sampling and purposive sampling and the choice made on particular occasions was arbitrary. The 1926 Report of the Commission set up by the International Statistical Institute to study the representative method described such samples as being "made up by purposive selection of groups of units which it is presumed will give the sample the same characteristics as the whole". The rigorous theory of design-based inference dates back to Neyman (1934), but it was definitively established as the dominant approach with the text-books written by Yates (1949) and Deming (1950) and even more assertively in those written by Hansen et al. (1953) and Cochran (1953). The properties of a design-based estimator are defined in terms of its behaviour *over repeated sampling* (i.e. in expectation over the set of all possible samples permitted by the sampling design).

During the early 1970s,however, the choice re-emerged and design-based or model-assisted sampling came to be questioned by Royall (1970) and consolidated by Royall and Herson (1973), who used purposive selection under another name. Studies reveal that both the approaches have their merits and disadvantages and the merits of both ca be suitably combined.

Suppose that the sampling strategy in a particular survey has to use stratified sampling and the classic ratio estimator. Within each stratum, a choice can be made between simple random sampling and simple balanced sampling. It can be shown that balanced sampling is better in terms of robustness and efficiency, but the randomized design has certain countervailing advantages.

The question that becomes relevant at this stage is how to obtain a 'balanced sample'. The definition suggested by Royall and his co-workers may not be easy to implement in all cases. That, of course, should not lead to an outright rejection of 'purposive selection' for the simple task of planning a sample survey that provides some idea about the population in respect of some features of interest.

5.2 Non-Probability Sampling

Probability-based sampling has been widely used and even accepted as the method of sampling that can yield valid statistical inferences about some large population (which can be theoretically canvassed in a census or a complete enumeration exercise) and provide quantitative ideas about the reliability of such inferences, However, recent concerns about coverage and non-response coupled with rising costs have motivated the use of non-probability sampling as an alternative method at least in some situations, specially in social science research. Non-probability sampling designs include convenience sampling, case-control studies and clinical studies, intercept survey design, opt-in panels and some others. Methods like network sampling use a mixture of probability-based and non-probability sampling. The report of the task force appointed by the American Association for Public Opinion Research (June 2013) provides a good review of non-probability sampling methods and their applications as well as their limitations and advantages. The Committee observed "Non-probability sampling may be appropriate for making statistical inferences, but the validity of the inferences rests on the appropriateness of the assumptions underlying the model and how deviations from those assumptions affect the specific estimates".

Two main approaches to conducting surveys were used earlier by official agencies, the first choosing a representative community (based on judgement) followed by a complete enumeration of that community and the other choosing sampling areas and units based on certain criteria (to take care of some co-variates) or controls akin to that of quota sampling. Neyman established the triumph of probability sampling of many small, unequal clusters, stratified for better representation (Kish, 1965).

In quota sampling a desired number of participants in the survey is attempted in a relatively short period of time, without the expense of call-backs, ignoring not-at-homes and refusals. Sudman (1967) used the phrase "probability sampling with quotas" to denote multi-stage sampling, possibly with stratification, where the first stage, second stage etc. units are selected with certain probabilities, but at the ultimate stage instead of referring to a list of all ultimate stage units and drawing a random sample therefrom, quotas of respondents according to their availability and according to interviewers' convenience are filled up. It is not possible to provide any estimate of sampling error in estimates from quota sampling or to set up confidence intervals for the population parameters of interest. If quotas from some specified characters like gender, age-group, educational qualification, etc. have to be attempted with targeted and not-small numbers for each specified group, the advantages of quota sampling over probability sampling would be lost. Theoretical aspects of quota sampling have been examined by Deville (1991).

In convenience sampling the ease of identifying, accessing and collecting information from potential participants is the determining consideration. Some common forms of convenience sampling are illustrated in mall-intercept surveys, volunteer surveys, river samples, observational studies and snowball sampling. In mall-intercept studies, every kth customer coming out of a shopping mall (through a particular gate) with a pre-determined k depending on the sample size targeted and available information about the flow of customers into the mall is interviewed. The time taken to fill the target sample size cannot be pre-specified and we can work out reasonable estimates about customers' behaviours and reactions. However, in some cases the practice is to select haphazardly some customers coming out – maybe successive ones – till the target or quota is filled up. Inferences based on this procedure are likely to be less reasonable than in the first case.

Probability sampling from a rare population, which could be a just a small fraction of the entire population, say of households or individuals, may be too costly when a complete listing of all units in the embedding population has to be carried out to identify units belonging to the rare population (or better sub-population) and may be challenging for example when the most of the units in the rare sub-population are pavement-dwellers or are nomad or homeless. In such cases, non-probability sampling may have to be used. Three alternatives are quota sampling, which may require an unpredictable time to complete the quota, targeted sampling and time-location sampling. Quota and targeted sampling are both purely non-probability. Targeted sampling is somewhat akin to network sampling. Time-location sampling (Mackellar, 2007) can be based on a probability sample drawn from a sampling frame having two levels, the upper level based on locations and times when members of the target group are known to assemble These time-locations are treated as strata from which either all units of the population or a probability sample is considered. It is somewhat similar to cluster sampling where the time-location groups serve the purpose of clusters. One limitation is the possibility of the upper level of the sampling frame not covering all units belonging to the rare sub-population.

5.3 Randomized Response Technique

Introduced by Warner (1964) and modified by Greenberg and others (1969), this technique has proved to be a useful sampling strategy seeking confidential or private information regarding sensitive or stigmatized issues where evasive responses (including non-response and biased responses) are common in the usual procedures to obtain information from the selected individuals or institutions. Over the last five decades immense methodological improvements have taken place in this technique and a lot of important theoretical research has been carried out and more is being carried out. However, not many substantive applications of this otherwise efficient strategy for information-gathering have been reported, except for some applications discussed in articles which offer some improvements over existing procedures or extensions to newer problems by way of illustrative examples.

Questions to which direct and truthful answers cannot be solicited or obtained may relate to tax evasion, substance abuse, support for some constitution amendment that is not acceptable to a large segment of the society, involvement in some criminal or immoral activities, etc.

Four basic designs for eliciting responses and subsequently deriving unbiased estimates of the proportion of 'yes' responses in the population sampled (assumed to be large enough) using simple random sampling with replacement have been generally discussed, viz. (1) the mirrored question design (also called the Warner RRT 1 design), where the idea to randomize is to know whether a respondent answers the sensitive question Q or its inverse Q', (2) the forced response design (introduced by Boruch, 1971), which is the most popular design, eventually having only a YES or NO response from each respondent, (3) the disguised response design (introduced by Kuk, 1990), in which the YES response to the sensitive question is to be identified with the YES stack of black and red cards and the response NO is to be identified with the NO stack, both stacks having the same number of cards and the same proportion of black cards in both, and (4) the unrelated question design.

Anticipating that including both affirmative and negative versions of the sensitive question may create confusion among the respondents, RRT or RRM was modified as RRM2 by introducing an unrelated question, such as "were you born in the first quarter of a year", denoted as Q*, which has two forms of true reply, YES or NO, the chance of a YES response being ¼. Hence the chance of a YES response to the sensitive question is now $P^* = pP + (1 - p) / 4$ and the estimate of P, the proportion of a YES response in the population, is

Est P = [f / n − (1 − p)/ 4] / p, with its variance given as

$$V \text{ [est P]} = [(1 − p)(3 + p)\, 8\, p\, (1 + p)\, P − 16\, p\, 2\, P^2] / 16\, n\, p^2$$

If the sensitive information item relates to a quantitative feature Y like income not stated in the income tax return or income earned through engagement not legally or otherwise permitted and we want to estimate the total or mean of Y in a finite population of size N, RRT has been applied conveniently when Y is taken to be discrete and the N values of Y are taken to be covered by a set of values like M_1, M_2, …. M_k. We may adopt SRSWR (n, N) or any other fixed sample size n design.

We prepare 25 identical cards. For each of five cards the instruction at the back reads: report your true income. For the remaining 20 cards, to be used in pairs, the instruction is to report any M_i = 1, 2, … 10, say. In this strategy, the chance of choosing a card with the instruction to report the true value of Y is 0.20 + δ say, while the chance of picking a card corresponding to a specified M_i is 2 / 25 = (1− δ) / K. Going by this scheme Y_i can be unbiasedly estimated as [R_i − (1 − δ) M*] / δ, where M* is the average of M values. Hence the mean of Y values has the estimate 5 R* − 4 M* under SRSWOR. For derivation of the sampling variance of this estimate and estimation of the same, one can refer to Mukherjee et al. (2018).

5.4 Panel Surveys

In many empirical research projects, we need repeated measurements on some features under study for units selected from the population of interest. Thus we may be required to assess the outcome of certain programmes and plans initiated by the state to improve maternal and child health by estimating some parameters related to several indicators of change/improvement in health. We have to conduct repeated measurements of these indicators on units selected from the population of children and mothers covered by the programme or plan. One obvious route is to select a panel or set of sampling units (households in this case) and to visit these households at specified intervals (annually or half-yearly or quarterly or monthly) to seek the relevant information to estimate both levels of the parameters for any specified visit or changes in any parameter during the interval between a specified visit and the earlier one. Similar is the case with the study of changes in productivity or gross value added per capita in manufacturing industry or among industrial units belonging to a specified sector of manufacturing.

If we consider fresh panels or samples of units with no overlap between successive surveys/visits (except by chance), we will not be able to estimate changes, though levels can be estimated conveniently. A balance between these two choices is effected through a rotational panel design, in which a part of the first panel is replaced by a new part during every repetition of the survey till the first panel is completely exhausted. This approach will reduce the variance of the estimated change compared to the case where we use different panels each time, since the covariance between successive visits to units in common will be positive. On the other hand, operational aspects like the apprehended burden on the respondents due to revisits and the effect of recalls should receive due attention.

Cantwell (2008) presents variances of different unbiased linear estimates (divided by the common variance σ^2) using a rotation design with 50% overlap and a composite estimator to get some idea about reduction in variances of level and difference estimates

Sample Design	Composite	Form	α	β	Var (level)	Var (diff)
No overlap	No	Simple average	0.5	0	0.5000	1.0000
50% overlap	No	Simple average	0.5	0	0.5000 0.6000	
50% overlap	No	α, β estimated for min var (estd. diff)	0.8333	0 .8333	0.7222	0.3333
50% overlap	Yes	k=2, α=.6, β=.1	0.6	0.1	0.4440	0.4800
50% overlap	Yes	α,β to minimize var estd. (diff.)	0.5952	0.2381	0.4048	0. 4240
50% overlap	Yes	Do	0.6565	0.2435	0. 4118	0. 4052
100% overlap	No	Simple average	0.5	0	0.5000	0.2000

5.5 Problems in Use of Stratified Sampling

5.5.1 Problem of Constructing Strata

In the previous chapter, we have illustrated the exercise of stratification in terms of choosing a stratification variable which is causally connected to the primary estimation variable and partition values for the chosen stratification variable to delineate different strata. In fact, the construction of strata in cases where such segments of the population of interest are rather evident or easily identifiable, say, by way of administrative divisions, natural differentiation by some auxiliary variable on which information can be easily collected etc. is a task by itself that defies unique solutions. Let us consider one such example in the same domain as the example in the previous chapter, viz. an establishment survey (covering only non-agricultural establishments with fewer than ten workers) to estimate the total number of workers engaged in eight identified sectors of industry which had been found to account for a vast majority of the total workforce in non-agricultural establishments of the size specified.

The frame available for this survey gave the names of such establishments in every village or town/city without providing the address. And for each establishment of size nine or fewer workers in every such first stage unit, the sector of industry to which the establishment belonged was also noted. Estimates of the total number of workers in each of the eight identified industry sectors were to be provided for each state, for rural and urban areas separately. For each of the two domains, viz. urban and rural areas, nine strata were formed in the following manner.

Stratum 1 S_1: to comprise all first-stage units (FSUs) having at least one establishment belonging to each of the eight industry sectors of interest.

Stratum 2 S_2: to consider all those remaining FSUs which have at least one establishment from any seven of the eight sectors of interest. It must be noted that none of the FSUs in S_2 will have establishments from all the eight sectors, However, some FSUs may not include any establishment from some sector which may be represented by at least establishment in some other FSU within S_2; thus, taken collectively, S_2 has a non-zero probability of representing all the eight sectors.

Stratum 3 S_3: to include all the remaining FSUs which have at least one establishment from any seven of the eight identified sectors of industry. This stratum also has a non-zero probability of representing all the eight sectors, though this probability is expected to be smaller than the corresponding probability for S_2.

In this way we proceed to define Stratum 8. We now add the last stratum S_9 as the aggregate of all the remaining FSUs. This stratum is not expected to show any establishment belonging to the eight sectors. Proceeding further, we can draw simple random samples without replacement from each of these nine strata within each domain separately.

5.5.2 Problem of Allocation of the Total Sample across Strata

The allocation n_h that minimises the sampling variance of the estimated mean

$$V = 1 / N^2 \Sigma N_h^2 S_h^2 [1 / n_h - 1 / N_h] \dots \tag{5.5.1}$$

where S_h^2 is the variance for stratum h taken as optimal. In the case of the same variance in each stratum, we have the proportional allocation given by $n h = n. N h / N$. If cost considerations are not taken into account, we have the optimal allocation due to Neyman (proved much earlier by Tchuprow) given by

$$n_h = (N_h S_h / \Sigma Ni Si) \times n \dots \dots \tag{5.5.2}$$

Given a cost constraint $\Sigma ni\, ci = C0$ where ci is the unit cost of a survey in stratum i, the allocation is given by

$$ni = [C0\, Wi\, Si / \sqrt{ci}] / [\Sigma Wi\, Si \sqrt{ci}] \dots \tag{5.5.3}$$

These formulae cannot ensure that all sub-sample sizes are integers and if the values yielded by any such formula are rounded off to near-by integers to make the sum n, there is no guarantee that the variance of the estimated mean will be minimized.

This is because we have not used any integer programming method to derive the optimal allocation. Secondly, formula (5.5.2) may result in (pathological) situations where $n h > N h$ as shown below

- stratum   1   2   3
- size   10 60 50
- SD   12 2   2
- n1 works out as 11 approximately, if we take n = 30.

Wright has given two algorithms which result in all integer sub-sample sizes which satisfy the constraint $\Sigma ni = n$. The algorithm can produce integer ni values which may be required to satisfy other constraints like ai (integer) $\leq$ ni $\leq$ bi (integer) for some i.

- In most cases, sub-sample sizes are required to satisfy some boundary conditions for cost or precision consideration like $n I \geq 2$ or $2 \leq ni \leq mi < Ni$, besides $\Sigma n I = n$. Sometimes, conditions like ni $\geq$ nj for some i and j may be relevant.
- An optimization problem where a solution is either an ordered sequence or a combination of decision variable values which satisfy some relations connecting them can be regarded as a combinatorial optimization problem.

- Thus the problem of optimal allocation is really a combinatorial problem and even with a moderately large population and a handful of strata, the number of candidate solutions may be quite large.
- The task of identifying the set of feasible solutions is not that easy.
- The problem is pretty difficult if we have multiple objectives to be achieved.
- When multiple characters are under observation, we have different SD values for different characters within each stratum. Several plans for overall allocation have been suggested in such cases. Some follow.

  (a) Minimize the sum of relative increase in the variances of the estimates, given by the overall optimal allocation, over variances of estimates from respective univariate optimal allocations Resulting sub-sample sizes are given by

  $$n_h = [C_0 \sqrt{\Sigma \, nlh * 2}] / [\Sigma \, ch \sqrt{\Sigma \, nlh * 2}]$$

  (b) Averaging the individual optimum allocations given earlier to yield

  $$nh = 1 / p \, \Sigma \, [C_0 \, Wh \, Slh / \sqrt{ch}] / [\Sigma \, Wh \, Slh \sqrt{ch}]$$

  (c) A compromise allocation can be worked out by using goal programming which sets an upper limit to the increase in variance of the estimate for the lth characteristic from the one using the optimum allocation for the particular characteristic. The objective function could be a weighted total of the increases.

- In a somewhat different approach, a pay-off matrix is constructed with rows corresponding to all possible allocations (n ,n_2, ... n_k) satisfying the conditions ni ≥ 1 and $\Sigma \, n_i = n$, columns corresponding to the different characters – or better the different parameters to be estimated – and the pay-off taken as the variance of the estimate for a particular character following a particular allocation.
- The optimum allocation corresponds to the minimax pay-off. Other optimality criteria like the minimax regret or the minimum expected pay-off could also be considered.
- One may even associate importance measures or weights to the different parameters and consider the weighted average of the variances for minimization.
- Any of the above objective functions will imply a complex formulation of the optimization problem, pose problems in identifying feasible solutions and involve a complicated discrete optimization problem to be solved numerically. In fact, we may have to develop some analogue of the branch-and-bound algorithm even to identify the different feasible solutions.
- While applications of optimization methods and techniques can lead to better and more defensible solutions to many vexing real-life problems, formulation of the corresponding optimization problems in terms of feasible solutions, constraints and objective function may not be unique and the usual robustness analysis does not reflect the inherent concern.

Most – if not all – sample surveys are multivariate in nature and we need to estimate several parameters, And in this context a big question, usually side-tracked, is the question relating to allocation of the total sample across the strata. Of course, proportional allocation does work even in multi-item surveys. However, when several parameters have to be estimated and we try to have somewhat like an 'optimal' allocation, with or without cost being incorporated in the objective function to be optimized, we run into the problem of handling the variances, or better, the variance-covariance matrix of the estimates. The elements in this matrix will depend on the allocation and also on the population variances and covariances of the several estimation variables of interest. Even if we presume that the latter elements are known or some reasonable prior values are available, we still confront the problem of choosing some scalar function of the matrix of variance-covariance of the parameter estimates and, then, minimizing that with respect to the vector of stratum sample sizes.

The first problem of choosing an objective function attracts concepts and techniques of A-, D- and E-optimality and also other optimality criteria used in the context of designs of experiments. This is quite an open area for research. There have been other approaches to determining the optimal allocation in multi-parameter surveys. For example, we can think of two procedures for estimating any single parameter based on an allocation based on this parameter alone and that based on the allocation to be worked out for the multi-parameter case, find out the corresponding variances and take the deviation between these two sampling variances. The 'optimal' allocation for the multi-parameter survey is then found by minimizing the weighted total of deviations (in the two estimates), of course, numerically. One cannot use any known algorithm for this optimization problem and recourse has to be had to some heuristic algorithm. Several researchers have worked on an allocation that will minimize the sum of squared coefficients of variation of the estimated means.

One of the earliest attempts to solve the allocation problem in multivariate surveys was by Folks and Antle (1965). The set of all possible allocations is a finite point set in p (number of estimation variables) $F = \{(n_1, n_2, \ldots n_k) \Sigma n_i = n, n_i$ an integer ≥ 0 for all I$\}$. The number of points in F is, as in the classic occupancy problem, by $(n - 1)! / [(p - 1)! (n - p)!]$. We like to estimate the population means of the p variables.

Let $y_i (n) = \Sigma N_h S_{ih}^2 / N^2 - \Sigma N_h^2 S_{ih}^2 / n_h N^2$. Then the point n^0 is better than n if $y_i (n^0) \geq y_i (n)$ for all I and $y_k (n^0) > y_k (n)$ for at least one k.

The feasible point n^0 is efficient if there exists no feasible point that is better than n^0. The set of all feasible efficient solutions is complete if, if given any feasible point n not in the set, there exists a point n^0 in the set that is better than n. Folks and Antle showed that the complete set of feasible efficient solutions is given by

$$N_h = ([\Sigma \alpha_i S_{ih}^2]^{1/2} N_h n) / [\Sigma N_h (\Sigma \alpha_i S_{ih}^2)^{1/2}$$

where S_{ih}^2 is the variance of the hth stratum for the ith estimation variable and the sums within parentheses are over the variables. Any solution from this set can be accepted in the sense that there will not exist a better allocation in the sense defined here.

5.6 Small-Area Estimation

In some sample surveys, we are required to produce estimates of certain parameter(s) of interest for segments of the population, not all of which have adequate representations in the sample that was selected without a consideration of this requirement. Otherwise, all such segments could have been recognized as strata to be represented more or less adequately in the overall sample. Thus, having adopted a sampling design like multistage stratified random sampling for a national or a state-level survey, we may take a post-stratification decision to estimate the parameter of interest separately for small domains like blocks or for deeply classified segments of the population classified according to sex × education level × occupational status. The consequence may be that in some such domains (or areas) of interest we have few sampling units/observations or even none. The usual methods of estimation would involve very high relative standard errors. Interest in small-area estimation problems has grown over the last 50 years and research workers continue to offer more efficient and robust estimation procedures. In fact, the demand for small-domain statistics (estimates) has grown rapidly and considerably where domains of interest cut across planned domains like strata in the sampling design adoptes for a survey and have low representations in the sample.

In 1967 the US bureau of Census initiated the Federal-State Co-operative Program to provide reliable and consistent estimates of county populations. The Household Survey Program of Statistics Canada provides monthly estimates of standard labour market indicators for small areas, such as federal electoral districts, census divisions and the Canada Employment Census through labour-force surveys. Small-area statistics are also useful in health administration, immigration records, educational surveys etc.

Besides the direct estimators based on values of the variable of interest only from the small area for which an estimate is to be offered, synthetic estimators and composite estimators along estimators making use of Bayes, empirical Bayes, hierarchical Bayes and empirical Bayes best linear unbiased prediction approaches have been discussed by a host of authors.

Estimates for a small domain could be taken as weighted combinations of a directly observed, unbiased estimate with a relatively large variance for the domain and a synthetic estimate with usually a small bias and a small variance, weights being so determined that the overall efficiency of these composite estimators (considering all such small domains) is maximized. The synthetic estimate could be some model-based estimator using auxiliary information or an unbiased estimate for a larger area.

5.7 Network Sampling

Link-tracing network sampling provides a useful strategy for leveraging network connections to facilitate sampling, specially in social or clinical research when the population of interest is rather small and a reliable frame of sampling units is not available. Even if a complete list of households in the area to be covered is available, it will not be worthwhile to visit quite a large number of households to identify households which have at least one member from the population of interest, e.g. individuals who are addicted to drugs/narcotics or are suffering from a disease which bears some stigma in society, given that the incidence of such addiction or disease is rare. In such cases, we can have recourse to link-tracing network sampling. The defining feature of this strategy is that subsequent sample members are selected from among the network contacts of sample members previously selected. In this way, network links are traced to enlarge the sample, leveraging the network structure to facilitate induction. With recent rapidly increasing on-line social networks, network sampling has become feasible on a large scale and is appealing in a practical and inexpensive way to reach a desired sample size.

In its early foundations, network sampling was not a non-probability sampling technique. Over the years, strict assumptions in probability sampling had to be relaxed in some situations and network sampling being used in such situations is now regarded as non-probability sampling. Goodman (1961) introduced a variant of link-tracing sampling called snowball sampling. In its original formulation, we assume a complete sampling frame, and the initial probability sample, viz. seeds, is drawn from this frame: k contacts of each seed are enrolled in the first stage (wave of the snowball), k contacts of each wave-one respondent form the second wave, and so on, until s waves are completed. Goodman advocated this sampling strategy for making inferences about network structures of various types within the full network.

Thompson and Cosmelli (2006) proposed several methods for sampling and inference using adaptive sampling strategies in which the sampling probabilities of different units are not known in advance but depend only on observed parts of the network. However, snowball sampling has come to imply even non-probability sampling where we may start with a convenience sample and include all contacts of each previous participant. Network sampling that starts with a random sample of seeds may yield reasonably good estimates of population parameters. The eventual sample size depending on the number of waves in snowball sampling may not admit of a simple probability distribution.

With two important modifications over the usual link-tracing sampling strategy, the respondent-driven strategy allows each initially selected participant to identify a specified number of contacts who would be passed coupons by the participant so that they could participate in the second wave, without any loss of confidentiality. Estimation of parameters and of associated uncertainty in this strategy remains a challenging task.

5.8 Estimation without Sampling

Over the last few years with a whole host of estimates of population characteristics or future behaviour patterns of individuals being called for by different public and private enterprises and agencies, sample surveys with associated costs and respondents' lack of interest and involvement in some cases are being replaced by amassing large volumes of data which are readily available as administrative records or in different communication media. In some situations, samples may be drawn usually systematically from the big data sets, thus collected quite conveniently. AAPOR has recognized three broad categories of techniques in this direction, viz. social media research, wisdom of crowds and big data analysis.

Google has demonstrated how aggregation of search data can be used to monitor trends in public perceptions about happenings that directly or indirectly influence their lives and also public behaviour. As mentioned in the AAPOR report, the central theme in wisdom of crowds is that a diverse group of informed and deciding individuals is better than experts in solving problems and predicting outcomes of current events like election results. It is argued that the estimate offered by each individual has some information and is subject to some error. Across the responding individuals the errors are assumed to yield a zero sum and the average of the estimates offered can be taken as a usable estimate of the outcome. This approach has been quite often used in predicting outcomes of elections.

Big data often imply huge volumes of data routinely and sometimes automatically generated for administrative, commercial or business purposes as well as from properly designed sample surveys. To derive reasonable estimates of labour force or employment in a particular sector like information technology or rate of unemployment in households and also establishment surveys are being planned and conducted by various official agencies. The quality of data collected and thus the estimates of any parameter of any interest arising from the survey are not always above board and are adversely affected by many factors associated with the behaviour of respondents, as also with the competence and performance of the investigators. In such a context, it may be argued that there exist large databases which are updated routinely for some administrative purposes like implementation of some labour laws. One such data set in India is maintained by the Employees' Provident Fund Organization (EPFO). Each organization with a minimum number of employees on its roll is mandated legally to put some information items regarding its employees and each employee has a unique identification number that takes care of issues related to migration of employees from one organization to another, also avoiding multiple inclusion of employees who serve more than one organization at the same time. Thus one may consider the EPFO figure as a credible estimate of the total number of employees in organizations with the specified minimum number of employees.

5.9 Combining Administrative Records with Survey Data

In several socio-politico-economic investigations data needed to build up models and to develop new theories or to validate existing ones may not be available entirely in administrative records and have to be augmented by survey data. Usually, surveys that are planned and conducted by official agencies and for which unit-level data are available in the public domain may not necessarily have adopted sampling designs to produce results at highly disaggregated levels which may be of interest to research workers. In such cases, careful pooling of survey data and relevant administrative records has been taken advantage of in several countries. For this purpose, post-stratification, raking and small-area estimation are three oft-used methods.

Thomson and Holmoy (1998) provide a comprehensive account of the Norwegian experience in combining survey data with administrative registers, with the primary objectives of (1) reduction in sampling variances of estimates, (2) reduction of effects of non-response and (3) consistency among estimates derived from different data sources. Post-stratification is the most widely used method. Both post-stratification and raking secure consistency between marginals published from various sources. Raking, however, cannot reduce sampling variance except marginally. Small-area estimation has been used quite profusely to produce estimates based on auxiliary information provided by administrative records, mostly available at the unit level.

Operational definitions of various concepts used in administrative registers often differ from corresponding definitions adopted for various surveys in many countries. This has been responsible for less than adequate use of administrative records to supplement information provided by surveys planned and conducted by national statistical offices in most of the countries. This has even affected any serious effort to improve the quality of administrative records, which have often been found to be lacking in accuracy, though claiming complete coverage We also have periodic (quinquennial or decennial) surveys on some aspects of the economy or society which also can be used in conjunction with related sample surveys.

6

More about Experimental Designs

6.1 Introduction

As noted earlier, experiments in different types of research on diverse phenomena have very different connotations in terms of experimental units, treatments applied, responses noted and principles to be followed in designing the experiments so that the objective(s) behind any experiment can be achieved as effectively and as efficiently as possible. A design in this context is effective if it can provide estimates (with some desirable properties like unbiasedness or minimum variance) of the parameters in which the experimenter is interested. And efficiency requires that resources required to get such estimates in terms of the number of runs should be as small as possible.

Quite often the focus is on optimization experiments and thus on response surface methodology. Interest in multiple responses may complicate matters. The concern for cost minimization – realized more in industrial rather than agricultural experiments – calls for some non-traditional designs. Even in biological experiments where ethical considerations dictate a small number of trials with living beings, we need special designs.

As pointed by Box, experimentation is essentially a sequential procedure where we may start with a few runs and a simple model and later augment the initial design by adopting a suitable strategy, instead of starting with a large number of runs.

6.2 Optimality of Designs

A lot of research has been reported and more remains to be taken up in the area of 'optimality' of experimental designs. Connotations of 'optimality' vary from one objective to another and, thus, from one type of model to another. The objective and the model depend on the type of experiments to be designed. In the context of screening experiments dealing essentially with the regression of yield or response on levels of several factors likely to affect yield, optimality refers to the choice of levels of the selected factors to ensure maximum efficiency of the estimates of regression parameters appearing in the model. Talking of comparative experiments and the ANOVA model, where treatment contrasts or factor effects have to be estimated, optimality refers to the estimates of contrasts and, consequently, of factorial effects and is to be achieved in terms of allocation of the given number of design points over the blocks or arrays and thus the type of the design. Coming to optimization experiments, we again have to estimate the optimal setting of design parameters with maximum efficiency. Here the concept of efficiency may have to imply certain desirable properties of the search for the optimal setting as well as of the estimated optimal setting itself. Speaking of comparative experiments, the inclusion of covariates brings in new research problems, many of which have been addressed by research workers. Sinha et al. (2016) have provided a detailed insight into several aspects of this problem.

Thus, the objective to be achieved could be to ensure efficient estimates of different effects (main and interaction), specially the effects of interest, with a minimum number of design points, by choosing the allocation of such points among the different treatment combinations. The objective could be just to minimize the number of design points subject to the requirement of providing estimates of parameters of interest with some specified efficiencies. Even for the first objective, we can think of different criteria, e.g. different scalars associated with the variance-covariance matrix of these estimates like the determinant or the trace or the maximum element or some other measure. Instead of going by the covariance matrix, we can think of a measure of distance between the vector of parameters of interest and the vector of the best linear unbiased estimators of these parameters. Here also different distance measures can be thought of.

The earliest theory of regression designs and related optimality results can be traced to Smith (1918) and Plackett and Burman (1946). Optimum allocation in the context of linear regression was first studied by Elfving (1952).The concept of continuous or approximate designs was introduced by Kiefer and Wolfowitz in 1959. D-optimality minimizing the generalized variance of estimates of the treatment contrasts and later D_s dealing with a subset of parameter estimates were discussed by several authors. D-optimality has the effect of minimizing the volume of the joint confidence ellipsoid for the estimated coefficients in the regression model. A-optimality minimizing the trace of the variance-covariance matrix of estimates and its extension by Fedorov (1972) and equivalence of some of these optimality criteria were also studied. A-optimality implies a minimum for the average prediction variance of the model coefficients. Kiefer (1975) unified the different optimality criteria in terms of the Φ-optimality criterion. Several investigators contributed to the development of algorithms to construct optimal regression designs. Sinha (1970) introduced the concept of DS-optimality in terms of minimizing the distance between the parameter vector and its best linear unbiased estimator in a stochastic sense. Properties of DS-optimality in discrete and continuous settings and certain generalizations were considered by Liski et al. (1999), Mandal et al. (2000), Liski and Zaigraev (2001) and a few others. One can think of a modified distance-based optimality criterion requiring that the model points have the maximum spread across the factor space and the D-optimality criterion is satisfied. An I-optimal design chooses runs that minimize the integral of the prediction variance across the factor space and hence is recommended for building response surface.

Research in the broad area of designs of experiments has proceeded on two distinct lines, viz. construction of designs with different properties and analysis of data collected through an experiment along with issues of choosing a design with some desiderata. While researches in the first direction are mostly deterministic in nature and have attracted the attention of pure mathematicians, the latter involve stochastic methods and thus have been enriched through contributions by statisticians also. Even industrial engineers have come up with different requirements of industrial experiments and have also offered designs that meet such requirements.

Complete text-books and reference books have been devoted to design and analysis of experiments and do not always cover the entire gamut of the problems arising in this context. The purpose of this chapter is essentially to discuss some relatively less treated – though pretty well-known – aspects of the subject and to indicate wherever possible some open problems for research. No attempt has been made to address problems connected with the construction of experimental designs.

Related to the question of optimality is the concern for robustness, which has engaged the attention of research workers since the late 1970s. Two distinct lines of work on robustness of designs are discernible. The first is to develop designs that guard against particular deficiencies like outliers, missing values, correlated observations, non-Gaussian distribution of observed responses and the like. In the second we characterize designs which are optimal relative to certain specific criteria as to how optimal they remain against other criteria. Andrews and Herzberg (1979) were interested in obtaining designs which are robust against loss of observations or outliers. Box and Draper (1975) wanted to minimize the effects of outliers on the analysis. Hedayat and John (1974), and John and Draper (1975) examined the robustness of balanced incomplete block designs. Herzberg et al. (1987) considered designs which retain equal information or ensure the same generalized variance for estimates of unknown parameters, regardless of which observations were lost. They showed that a balanced incomplete block design with five treatments in ten blocks of size three each ensures equal loss of information with respect to any single missing observation or any pair of missing observations, but not any triple. In fact, robustness in the context of experimental designs and in respect of various likely disturbances still fascinates research workers.

6.3 Fractional Factorial Experiments

A full factorial experiment, say with 15 factors each at two levels, will require 2^{15} experimental runs, that too with only one replicate. This number is quite high, specially in situations where a run of the experiment involves a lot of time and money. In such cases we go for fractional factorial designs, foregoing information about unimportant interactions. Orthogonal arrays have been used along with linear graphs in fractional factorial designs. Orthogonal array designs were vigorously pursued by G. Taguchi in the context of industrial experimentation. An $N \times n$ array of s symbols is said to be an orthogonal array (OA) of strength t if every $N \times t$ sub-array contains every t-plet of the s symbols an equal number of times, say λ. This OA is usually denoted as OA (N, n, s, t) where N stands for the number of experimental runs, n denotes the number of factors and s denotes the number of levels of each factor.

It should be remembered that in a screening experiment, our interest lies in identifying factors which affect the observed responses(s) significantly. Hence we have to get estimates of the main effects of these factors suggested from domain knowledge. Estimation of interaction effects is not that important and we can afford to lose information on interactions to minimize the number of runs. The best strategy for screening experiments is to choose a subset or fraction of the experimental runs required for a full 2^k design in all the factors under study. Two types of screening designs, both obtained by taking a subset of the full 2^k factorial design, which are commonly used are: (a) fractional factorial designs and (b) Plackett-Burman designs. The fractional factorial designs are obtained by taking a regular fraction of a 2^k design like ½, ¼, ⅛ (i.e. where the denominator is power of 2). For example, to study the effect of six factors a ½ fractional factorial would consist of ½ ×

$2^6 = 32$ runs. A ¼ fractional factorial would consist of ¼ × $2^6 = 16$ runs. The Plackett-Burman designs are constructed in increments of four runs. So, for example, you could study six factors using a Plackett-Burman design with only 12 runs. It must also be noted that whenever we try to minimize the number of runs, we invite and accept loss of some information.

In a fractional factorial design, we choose a subset of runs from a full factorial without confounding main effects. In constructing half replicates of 2^k designs, only half the total experimental runs are made, therefore each factor effect and interaction that could be estimated had one other interaction effect confounded with it (they are confounded in groups of two). If one-quarter of the total experimental runs were made, each factor effect and interaction would have three other interactions confounded with it (they would be confounded in groups of four). If one-eighth of the total experimental runs were made, the factors and interactions would be confounded in groups of eight, and so on – the greater the fraction, the greater the confounding. The interactions between the factors do not go away just because we do fewer experiments; instead, their impacts cannot be separately estimated.

To construct a one-quarter replicate, we start with a full factorial in k-2 factors, called the base design. This will give us the number of runs we would like to have. Now, we have two factors left over whose levels still must be set. These *added factors* are taken care of by associating (confounding) them with two of the columns of the calculation matrix. For example, in constructing one-quarter replicate of a 2^5 factorial, we start by writing a 2^3 based sign, then we associate the added factors, 4 and 5, with two interactions in the calculation matrix. Let us arbitrarily use the 12 and 13 interactions for s_4 and X, respectively. That is, our generators are: 4 = 12 and 5 = 13.

From the generators we find two of the interactions confounded with the mean, which are 1 = 124 = 135. The third interaction can be found by multiplication, recognizing that if 124 and 135 are both columns of $^n+^n$ signs (= 1), then when we multiply them together we still get a column of $^n+^n$ signs. Therefore, the last interaction is 1 = (124)(135) = 2345, and the complete defining relation is 1 = 124 = 135 = 2345.

Next the entire confounding pattern can be obtained by multiplication as before, so that

1	=	24	=	35	=	12345
2	=	14	=	1235	=	345
3	=	1234	=	15	=	245
4	=	12	=	1345	=	235
5	=	1245	=	13	=	234
23	=	134	=	125	=	45
123	=	34	=	145	=	25

Note that the first column corresponds to the contrasts we can estimate, which we obtain from the headings of the columns in the calculation matrix.

When we construct higher fractional replicates like 1/8 or 1/16, we will have the mean confounded with 7 or 15 other interactions, respectively. For the 1/8 replicate, we start by writing down the full factorial in k-3 factors and then associating the last three factors with any three interactions of the first k-3 factors (i.e. we need three generators). This, in effect, specifies three interactions that will be confounded with the mean, 1. Then, the other four interactions confounded with the mean will be all of the possible pair-wise products, and the three-way product of the original three interactions in the defining relation. For the

1/16 replicate, it is necessary to specify four generators which will give us four interactions in the defining relation. Then, the remaining eleven interactions will be all of the possible pair-wise products, three-way products, and the four-way product of the original four.

Since there are many ways of generating a regular fraction of a 2^k factorial design, it is useful to have a measure of how 'good' a particular design (or fraction) is. One measure that is used is called *resolution*. Resolution is defined as the length of the shortest word in the defining relation. For a half fraction there is only one word in the defining relation (other than the letter I). For example, consider the half fraction of a 2^5 factorial design with the generator 5 = 1234. The defining relation is I = 12345, Since the shortest (only) word (12345) has five factors in it, this resolution is five.

Because a resolution V design has main effects and two factor interactions confounded only with three factor interactions and higher-order interactions, these designs have often been adopted for industrial experiments.

As an alternative way to avoid the problem of confounding the (main) effects of factors under study, Plackett and Burman in 1946 developed a set of tables to be used for screening designs. Their tables are used to construct resolution III designs.

A full factorial in seven factors would take 2^7 = 128 experiments or runs. The Plackett-Burman design allows seven factors to be examined in only eight runs. This is a huge reduction in the amount of experimentation, but it does not come without a penalty.

The price we pay is that we cannot estimate any of the interactions. This is a resolution III design and they are confounded with the main effects of the factors.

The information we want from our data requires that:

$$n_F \geq k + 5 \text{ (for information)}$$

A class of designs which is more frugal with experiments is called *central composite* designs. Central composite designs build upon the two-level factorial design that is discussed in chapter 4. Remember that the model used to fit the data from 2^k-factorial design was of the form:

$$\widehat{Y} = b_0 + \sum_{i=1}^{k} b_1 X_1 + \sum_{i-1}^{k-1} \sum_{j=i+1}^{k} b_y X_i X_j \tag{1}$$

with interaction terms of higher order usually neglected. Equation 1 reminds us that the 2^k-factorial design allows the estimation of all main effects and two-factor interactions. The only terms missing to give us a full quadratic equation are the squared terms in each X_i. In order to permit the estimation of these terms, the central composite design adds a set of axial points (called *star points*) and some (more) centre points. The axial points combined with the centre points are essentially a set of one-at-a-time experiments, with three levels of each of the independent variables, denoted by $-\alpha$, 0 and α (where α is the distance from the origin to the axial points in coded units). With the three levels of each X_i, the quadratic coefficients can be obtained.

It should be noticed from the figures that α is not 1,0, so that the star points extend out beyond the cube of the factorial points. The only things to decide are how large α should be and how many replicated centre points should be included in the design. To answer these questions, statisticians have brought in two additional criteria; *rotatability* and *uniform precision*. Rotatability implies that the accuracy of predictions from the quadratic equation only depends on how far away from the origin the point is, not the direction. This criterion fixes α. The other criterion, uniform precision, means that the variance of predictions

TABLE 6.1

Number of Runs for a Central Composite Design

Number of Factors, k	No. of Runs in Central Composite Design (Factorial + Star + Centre)	No. of Coefficients in Quadratic Equation
2	$2^2 + 4 + 6 = 14$ ($\alpha = 1.41$; two blocks)	6
3	$2^3 + 6 + 6 = 20$ ($\alpha = 1.68$; three blocks)	10
4	$2^4 + 8 + 6 = 30$ ($\alpha = 2.00$; three blocks)	15
5	$2^4 + 10 + 7 = 33$ ($\alpha = 2.00$; two blocks)	21
6	$2^5 + 12 + 10 = 54$ ($\alpha = 2.38$; three blocks)	28
7	$2^6 + 14 + 12 = 90$ ($\alpha = 2.83$; nine blocks)	36

should be as small in the middle of the design as it is around the periphery. This fixes the number of centre points.

The value of α for all the designs was actually calculated using the formula: $\alpha = \sqrt[4]{\text{number of factorial points}}$.

All the designs can be run in blocks. The factorial portion with centre points constitutes one or more of the blocks, and the star points with some more centre points is another block. This gives two big advantages to the design. First, the factorial portion can be run *and analysed* first. If curvature is found to be negligible, then the star points need not be run at all (a big saving). Secondly, blocking is an important tool whose main purpose is to increase the precision of the results.

Another class of commonly used designs for full response surface estimation are called Box-Behnken designs, and they have two advantages over the central composite designs.

First of all, they are more sparing in the use of runs, particularly for the very common three- and four-factor designs.

The second advantage is that the Box-Behnken designs are only three-level designs (i.e. each factor is controlled at only −1, 0 or +1, or +α over the course of the set of experiments). On the surface this may seem to be totally inconsequential. But in an industrial setting, keeping the number of levels down to the bare minimum can often be a great help in the practical administration of an experimental programme.

It should also be stated that Box-Behnken designs also meet the criteria of rotatability (or nearly do) and uniform precision of predictions, and can be broken down into two or more blocks.

6.4 Other Designs to Minimize the Number of Design Points

Central composite designs are best suited to explore response surfaces in optimization experiments with a small number of runs. They build upon two-level factorial designs to estimate parameters in the quadratic response model used in 2^k-factorial designs, viz.

$$Y = b_0 + \Sigma\, b_i\, X_i + \Sigma_i\, \Sigma j\, b_{ij}\, X_i\, X_j$$

with second- and higher-order interactions usually neglected. However, we are able to estimate all the main effects and two-factor interactions in terms of the coefficients in this equation. The only terms missing here to explore a full quadratic response equation are the squared terms in each X_i. To accommodate estimation of these terms, we add a set of axial points (or star points) and some (more) centre points. The axial points along with the central points are essentially a set of one-at-a-time experiments, with three levels of each of the independent variables, denoted commonly by codes $-\alpha$, 0, α (where α is the distance from the origin of the factor space to the axial points in coded units). With these three levels of each X_i, the quadratic coefficients can be obtained.

6.5 Mixture Experiments

Mixture experiments are quite common in food-processing, pharmaceutical, chemical and textile industries where different ingredients have to be mixed in proportions to be optimally determined so that the product formulation can meet some desired specifications. In some cases, we deal with some factors for which a few discrete levels are considered, along with some continuously varying factors.

In agriculture, a fixed quantity of inputs like fertilizer, insecticide or pesticide is often applied as a mixture of several components. Thus crop yield becomes a function of the mixing proportions. An important application lies in the area of inter-cropping in dryland farming: the component crop is introduced by replacing a part of the main crop and the mean response per unit area will depend on the size of this part. In the food-processing industry, one tries out different fruit pulps or juices in different proportions to offer a product that has the desired taste or smell or colour or all these.

In such a mixture experiment, which is usually an optimization experiment dealing with continuously varying factors, viz. proportions, the factor space is a simplex and usually a collection of lattice points within that. Three common types of mixture design are

1. simplex lattice designs
2. simplex centroid designs and
3. axial designs.

A simplex lattice design consists of all feasible combinations of the mixing proportions, wherein each proportion takes the values $(0, 1/m, 2/m, 3/m \ldots m/m = 1)$ for an integer m. There are $(m+1)^q$ possible combinations of which only those combinations $(x_1, x_2, \ldots x_q)$ are feasible in which the sum of these elements equals unity. The set of design points is called a (q,m) simplex lattice and the number of points here is $C(q + m - 1)$, where $C(a,b)$ stands for the usual binomial coefficients involving positive integers $a \geq b > 0$. For example, a (4,3) simplex lattice has 20 design points.

A simplex centroid design deals exclusively with the centroids of the coordinate system, starting with exactly one non-zero component in the mixture (having q centroid points) and extending up to q non-zero components (having the unique centroid point, viz. $(1/q, 1/q, \ldots 1/q)$ With q = 4, there are 15 design points in this case.

Axial designs contain interior points on the axis joining the points $x_i = 0$, $x_j = 1/(q-1)$ and $x_i = 1$, $x_j = 0$ for all I not equal to j.

An exact design deals with an integer number of replications of the design points, thereby resulting in a design with an exact integer number of observations. For example [(0.4, 0.3, 0.2, 0.1) (3); (0.2, 0.4, 0.1, 0.3) (4); (0.1, 0.1, 0.4, 0.4) (2); (0.1, 0.2, 0.2, 0.5) (4)] produces an exact design with four distinct design points, with repeat numbers 3, 4, 2, 4 respectively, so that 13 observations are produced upon its application. In an approximate design, only the mass distribution over the design points is specified and, for a given total number of runs, the numbers of replications for the different design points are obtained approximately, in some cases.

Estimation of parameters in the mixture model

$$\Phi(x) = \Sigma \beta_i x_i \qquad \text{Linear or}$$
$$= \Sigma \beta_i x_i + \Sigma\Sigma \beta_{ij} x_i x_j \quad \text{Quadratic or}$$
$$\text{a cubic or special cubic (mean) response function}$$

remains the bedrock of analysis here. Some non-standard mixture models have also been suggested in some special situations. However, building up the response surface (with a non-linear response function) and determining the optimizing proportions has been the main task in this context. Estimability of the parameters and the information matrix and also optimality criteria have been studied quite comprehensively.

Example 1
An inter-cropping experiment was conducted by Dr. Panjabrao Deshmukh Krishi Vidyapeeth, Drylan Research Station, Akola, on redgram and safflower with the objective of finding the optimum proportions of area to redgram and safflower so as to maximize the gross returns. There are five inter-cropping treatment combinations ITCs defined as

ITC 1	Sole redgram
ITC 2	Redgram + safflower in the ratio of 2:1
ITC 3	Redgram + safflower in the ratio of 1:2
ITC 4	Redgram + safflower in the row ratio of 1:1
ITC 5	Only safflower

Thus proportions of area allocated to redgram and safflower are 1:0, 2:1, 1:2, 1:1 and 0:1 The study revealed that the optimum proportions of redgram for different price-ratios came out as [1.0: 0 .75; 24.49] [1.0:0.8;20.20] [1:1; 11.15] [1: 1.5; 3.9] [1: 1.75; 2.41] (Dekhale et al. 2003).

Example 2
Spillers Milling Ltd. in the UK conducted an experiment to study the effect of mixing four different wheat varieties in producing a good-quality bread. Specific volume (ml / 100 mg) was the response to be noted. Possible mixing proportions taken were 0, 0.25, 0.50 and 0.75 and 13 design points and 4 blocks of experiments in which 5 combinations, viz. (0, 0.25, 0, 0.75), (0.25, 0, 0.75, 0), (0, 0.75, 0, 0.25), (0.75, 0, 0.25, 0) and (0.25, 0.25, 0.25, 0.25), were common to all blocks, and Blocks 1 and 3 had four other points and four different points were present in Blocks 2 and 4. were carried out. A quadratic response function was fitted and the coefficients were tested for significance. Both the fitted regression equations keeping all the coefficients ands also keeping only the significant ones were considered. It was found that the maximum response of 438.9 came out for the combination (0.283, 0.717, 0, 0)

In many industrial research and development exercises, components to be mixed are required to satisfy some constraints like some upper limits not to be exceeded or lower limits to be necessarily exceeded or to lie between two specified limits or to meet some inequalities connecting the component values or proportions. An elaborate discussion on constrained mixture experiments which can be handled by introducing some pseudo-components or by using the extreme vertices designs is contained in Lawson and Erjavec (2001).

6.6 Sequential Experiments: Alternatives to Factorial Experiments

Many multi-factor experiments are meant to simply find the level-combination that corresponds to the optimum response at least approximately. Information about the main and interaction effects may not be that important and we are able to find the optimum level combination in as few runs of the experiment as possible.

Consider the metallurgical experiment to find the best composition of a certain type of alloy for maximizing the time to rupture under a given stress at a given elevated temperature. Location of the maximum is of primary importance. However, mapping the response function in the proximity of a maximum is important, since it may not be possible to control the factor levels as closely in production as in a laboratory.

Deficiencies of a factorial design in this context are:

It generates observations (on the response variable) to explore regions that may turn out in the light of the results, to be of no interest because they are far away from the optimum.

It explores a given region of the response surface comprehensively, and so can explore a either small region or a large region superficially.

If a small region is explored, it may turn out to contain no local optimum. The experiment could simply indicate the direction in which to move the experiment. If a large region has been covered with distant level combinations, an optimum may be missed entirely or, if found, will be found imperfectly so that an additional experiment is required.

In such cases, a sequential design in terms of the following steps is preferred.

Step 1. Select some particular factor level combination, which should be the best advance estimate of the optimum.

Step 2. Arrange the factors in some order.

Step 3. Have experiments (runs) with this combination and a few levels of factor.

Find the level of the first factor that gives the optimum response, levels of other factors kept fixed at the initial. Fit a polynomial, if necessary, to find the value of the first factor yielding the optimum.

Step 4. Hold the first factor at this value and the other factors at the initial levels, vary the second factor over a few levels and repeat Step 3 with the second factor.

Repeat this process with the third, fourth and remaining factors.

Step 5. After all factors have been varied, the whole process should be repeated, except that in Step 1 we start with the optimizing level of Factor one. This will provide the necessary number of rounds or replications. The rate at which the overall optimum response changes will indicate broadly the number of rounds needed to locate the optimum combination.

With n factors each at k levels, a full factorial design will require k^n observations in each round.

A single round of the sequential design will require k n design points or observations.

A single round of the full factorial design will thus permit $k^n / k n = k^{n-1} / n$ rounds of the sequential design, where the number of rounds is not known in advance.

This number does not depend sharply on n and depends primarily on (1) the closeness of the original starting point to the optimum combination and (2) the size of the experimental error relative to the differences in response considered to be important for detection.

Consider a $5 \times 3 \times 3 \times 3$ design which requires 135 design points or generates 135 observations in one round. A single round of the sequential design requires 14 points. Thus one round of the full factorial will permit around 10 rounds/replications of the sequential experiment. Fewer than 10 rounds may suffice if the initial point chosen is close to the optimum.

6.7 Multi-Response Experiments

- The goal of multi-response experiments is to find the setting of the design variables that achieve an optimal compromise of the response variables. By 'optimal compromise' we mean finding the operating levels of the design variables such that each response characteristic is as 'close' as possible to its ideal value.

- An alternative analysis procedure could combine the several responses into a single one which can then be optimized.

- We note that different response variables

  may be expressed in different units
  may be inter-related to different extents
  may justify different weights or importance measures in the context of the experiment.

- To get over the first problem, we can think of appropriate dimension-less transforms of the response variables Y1, Y2, ... Yk.

Dividing by the respective SDs is one option.

The other option is to bring in the concept of a desirability function, which takes into account the target and the tolerable values of a response variable. Different forms of the desirability function are available.

- Desirability is a distance metric ranging in value between 0 and 1, a value closer to 1 being more desirable.

- The distance metric may have a second component, viz. variance, as an extension of Taguchi's loss function.

- Variance of the predicted response may also be used to normalize the squared distance from the target.

Case 1 Target (T) is the best, the objective is to minimize w.r.t. x $[y * (x) - T]^2$, where $y * (x)$ is the estimated response y using estimates of coefficients in the regression equation of y on x (the vector of factorial effects). In this case, the desirability associated with an observed y is

$$d\,(y) = 0 \qquad\qquad\qquad\quad \text{if } y*(x) < L$$
$$[(y*(x)\text{-}T)/(T\text{-}L)]r \quad \text{if } L < y*(x) < T$$
$$[(y*(x)\text{-}U)/(T\text{-}U)]r \quad \text{if } T < y*(x) < U$$
$$0 \qquad\qquad\qquad\qquad \text{if } y*(x) > U$$

L and U are lower and upper acceptable values of y, while T is the target value. The parameter r determines the shape of the desirability curve. A value r = 1 gives a linear relation, while r > 1 yields a convex curve implying greater importance to values close to the target and 0 <r < 1 means concavity.

Desirabilities of different responses are usually combined by taking the geometric mean of individual desirabilities, viz. D = (d1d2 … dm)1 / m.

- If 'smaller is better', we have

$$d\,(y*) = 1 \qquad\qquad\qquad\quad \text{if } y*(x) < L$$
$$[(y*(x)\text{-}U)/(T\text{-}U)]r \quad \text{if } T < y*(x) < U$$
$$0 \qquad\qquad\qquad\qquad \text{if } y*(x) > U$$

- If 'larger is better', we have

$$d\,(y*) = 0 \qquad\qquad\qquad\quad \text{if } y*(x) < L$$
$$[(y*(x)\text{-}L)/(T\text{-}L)]r \quad \text{if } T < y*(x) < T$$
$$1 \qquad\qquad\qquad\qquad \text{if } y*(x) > T$$

- In a three-factor experiment using a central composite design, three response variables were noted and the following cubic regression was fitted to each response.
- The predicted maximum values of the responses were 95.38%, 76.17% and 81.97% respectively, with individual desirability values as d1 = 1.0, d2 = 0.69 and d3 = 0.70.
- Going by the overall desirability, the maximum value achieved was D = 0.78 at x 2 = 46.84, x 3 = 6.77.
- The polynomial fitted by least squares to yj

$$yj = \beta 0 + \beta 1 \times 1 + \beta 2 \times 2 + \beta 3 \times 3 + \beta 4 \times 1 \times 2 + \beta 5 \times 1 \times 3 +$$
$$\beta 6 \times 2 \times 3 + \beta 7 \times 12 + \beta 8 \times 22 + \beta 9 \times 32 +$$
$$+ \beta 10 \times 1 \times 2 \times 3 + \text{\euro}$$

The coefficient of x1 came out to be insignificant, as was the product term x2 x3.

The estimated regressions, omitting terms in x1 and x2 x3, and the corresponding R2 values came out as 0.8777.

6.8 Design Augmentation

The idea of experimentation, particularly in the industrial research and development context, being essentially a sequential process was first introduced by Box and Wilson (1951).

Box and Hunter (1957) made key contributions to response surface designs and response exploration to find the optimum treatment combination. The usual method of starting with a Plackett-Burman design or a central composite design followed by applying the steepest ascent (SA) method requires a lot of runs of the experiment and thus resources. The method of steepest ascent may be effective or not, depending on the nature of the underlying response model, the magnitude of experimental errors and the number of runs available.

Several authors have discussed design augmentation in a fixed design space to improve or repair model estimation, to de-alias model terms and/or to identify design points at which to conduct model validation tests. However, the issue of a two-phase design allowing a shift in the design space was first proposed by Nachtscheim and Jones. Nachtscheim and Jones (2018) have shown that an optimal augmentation strategy that places design points in a new region surrounding the SA direction can be more effective for both model estimation and response optimization, with the same number of runs used in one go.

Nachtscheim and Jones assume that (1) the underlying response surface is quadratic and the experimental errors are IID, normal, with mean zero, (2) with initial screening experiments performed, only two or three active factors remain in the follow-up experimentation, (3) the search for an optimum will not extend beyond two scaled units from the centre of any factor's range, (4) augmentation of design points is logistically feasible and there is no constraint to move to a new experimental region and (5) only one response is to be maximized. The authors suggest four different augmentation strategies with two initial designs, viz. the Plackett-Burman design and the definitive screening design, and two strategies to collect additional data.

The strategies considered by the authors are essentially guided by the works of Dykstra (1971) and Hebble and Mitchell (1972) and depend on the following. Mee (2002) provides extensions.:

(a) The order of the phase-one design. This could be a first-order orthogonal screening design or a definitive screening design (DSD). The latter can provide some estimate of the second-order response surface in phase one.

(b) The direction of shift in the design space. If a DSD has been used in phase one, the shift will be in the direction of the optimization trace, while if a first-order design is used in phase one, the shift will be in the direction of steepest ascent.

(c) The magnitude of the shift in the design space centre to be usually between 1.5 to 2.0 coded units.

(d) The shape of the design space, which could be either a hyper-cube or constructed from a set of candidate points which populate either the confidence cone of the SA direction or a simulated confidence region for the optimization trace.

Based on extensive simulation experiments with various combinations of first-phase and second-phase designs. Nachtscheim and Jones conclude that (1) an optimal design augmentation strategy is preferable to the method of steepest ascent in a sequential design strategy and (2) if only two or three factors are indicated as active on the basis of initial screening, using a DSD for the initial experiment is preferable to the standard two-level orthogonal array for finding an optimal factor setting in a subsequent augmentation experiment. They also point out the need for further research in this area.

6.9 Designs for Clinical Trials

Clinical trials are designed to establish the relationship between dosage and toxicity of new drugs for chemotherapy (Clinical Trials in Cancer Phase I) or to compare the efficacy of different treatment protocols and procedures. We may have to deal with quantal responses like cured and not-cured or with quantitative responses like TWIST (time without toxicity and symptoms) or the maximum tolerated dose (MTD). This is defined as the dose corresponding to an upper quantile of the distribution of tolerance, often taken to be logistic or normal. For planning such trials, one requires exact knowledge about the dose-response relationship. Since this relationship is scarcely known, if at all, adaptive designs are conducted in a stage-wise manner on individuals or groups of patients, so that the information gathered in intermediate stages can be utilized to approach optimal levels. Adaptive designs have been discussed in detail by Rosenberger (1996), Zacks et al. (1997), Biswas (2001) and others.

7

Models and Modelling

7.1 The Need for Models

Research workers use a wide variety of models to represent some real-world phenomenon or process or system (to be generally taken as an entity) to facilitate its study. In fact, models are simplified (sometimes even simplistic) representations of reality that allow incorporation of changes conveniently to investigate cause-effect (input-output) relations operating on the real-life entity. Responding to the need to represent different types of phenomena or systems for different purposes, an almost infinitely large plethora of models have been developed and used by research workers. And a wise choice from this huge stock to be used in a given context is by itself a research problem. At the same time, as new phenomena are investigated, the need for new models is often felt and research just to develop such models and to examine their applicability to deal with certain phenomena and, if found suitable, to derive useful knowledge about the underlying phenomena in modelling grows.

Models are required to

Describe the entity under consideration in all its relevant details, including its dynamics and interactions. In attempting to work out a description that will be comprehensive enough to enable some of the following activities, we have to identify and incorporate in the model all variables which can be controlled by us, those which lie beyond our control and are already specified and those which will be observed as the model is implemented in practice, as well as all constraints like equation or inequalities or other relations that connect the different types of variables. In addition, when some variables are found to be affected by unpredictable chance causes and behave randomly, we have to work out or assume appropriate probability distributions.

Analyse the entity by establishing the cause-effect relations governing the entity. This is the most important task for which models are used. One simple and common task here is to identify and to quantify dependence relations (including two-way ones) among the variables used in describing the phenomenon or process. In fact, it is expected that the consequence variables which are observed should depend on the controllable decision variables, conditionally given the values or levels of the given or constraining variables and relations. Even before proceeding with this task, we may like to examine possible grouping of the individual observations on the basis of their relative distances or to find suitable linear compounds of the variables which can explain their variances and the like.

Predict the state or behaviour of the entity at a future point of time or during a future period of time, assuming factors affecting the behaviour will continue to remain at their current levels or assuming some changes in these levels.

Control the state or behaviour of the entity at some preferred level by maintaining levels of the above-mentioned factors at some desired levels, and

Optimize (in some cases) the state or behaviour by deciding on the levels of controllable factors operating on the entity taking cognizance of some constraints on such variables and also on parameters linking these factors with the behaviour of the system or process.

Models are also involved in the premises of inductive inferences. And inductive inferences pervade the field of evidence-based or empirical research. In this context, a model is a set of assumptions made to process the evidences (data) for reaching some conclusion(s). Thus, probability models are used to represent the pattern of variation among the data units or observations, not simply reflecting the pattern observed in the finite data set but by recognizing the nature of the variable(s), the mechanism of data collection, the findings of similar studies carried out earlier and the like. In fact, probability distributions or models are the models most often spoken of in the context of statistical research. Dependence of some variable(s) on some other(s) or inter-dependence among the variables in the system has also to be modelled and incorporated in the premises for any inductive inference.

Models representing observed phenomena linked to farming and the farmers that go beyond the traditional models for physical aspects of farming to account for human behaviour aspects that are influenced by many external human interventions are assuming greater and greater significance. Appropriate models have to be developed – maybe through necessary modifications – and solved towards this. Models to explain changes in land use for different purposes – agricultural, residential, industrial, recreational, commercial etc. – and also in use of cultivable land to be put under different crops or crop rotations have sometimes been more qualitative but provide deep insights into social, cultural, political and economic systems and their operations.

While most of the above-mentioned models have a wide applicability in many different systems and processes, some other models which often admit of simple graphical presentations have been proposed in specific contexts. These models are being modified and refined as more evidence accumulates on the systems or processes modelled. It is also possible that these models may apply in some different contexts in which some research workers are engaged. At least, these models can provide some inputs for developing models that suit the systems or processes under consideration at present. We just take two such models for the sake of illustration. One relates to the process of software development and the other is in the context of knowledge management.

Starting with the early unstructured code-and-fix procedures followed mostly by individuals to develop software to the modern-day team work to develop software to carry out a whole host of functions – some identified and demanded by users and some others by the development team – within a reasonable time, different models for the development process have been suggested. The waterfall model brought much discipline to the process, though accomplished through rigidity. Problems with this model led to other models like the evolutionary and the transformation model. Thus the spiral model is a meta model since it can accommodate any development process model. An assessment of these models reveals that each one has some limitations and no one can be taken as applicable to all organizations for developing all types of software one for all. In fact, the realization is that before starting the process of developing new software (usually system software) the team should first come up with a process model that can derive bits and pieces from existing models and injects other features suiting the particular context.

7.2 Modelling Exercise

Modelling is the generic name for the exercise that involves model-building, model-testing and model-solving. The first task is to either develop a model that provides an effective representation of the underlying entity or to choose some existing model for the purpose as appears to provide an effective representation or has provided a reasonably good representation of a similar entity in the past. Building up a model to represent an underlying phenomenon or system, starting with some postulates or with some assumptions about the behaviour of the underlying entity, often supported by empirical evidences, has been a major activity with research workers. In fact, a lot of research has been continuously going on to come up with new models to represent new entities, taking cognizance of some properties which are not always overt or have not received due attention earlier.

This activity is followed by testing the suitability or effectiveness of the model either built up by the research worker or selected from a kit of existing and relevant models. The model has to be adequate in terms of taking due account of the cause system operating on the entity and should provide an effective answer to the problem of prediction or control or optimization as may be involved in the research. Beyond testing the relevance and/or adequacy of a model, we sometimes try to validate the model using some test data set and using some definition of validity. Concurrent validation against some external criterion is often preferred. Predictive validity is undoubtedly important, though this form of validation implies some prospective result.

Model-solving implies different activities in different contexts. Thus, solving a deterministic model will usually yield the value of some dependent variable or predictand given values of the independent or predictor variables. It may also yield the value of an unknown variable of interest given values of the parameters in an equation. Solving a stochastic model like a queueing model yields a discrete probability distribution, usually of the number of units present in the queue – either time-dependent (in which case we have a distribution with a probability density function depending on time t) or stationary (independent of the time of observation). On the other hand solving an inventory model generally yields an optimal order quantity. Solving a regression model really implies estimation of the regression coefficients from the observed data and using the same for prediction or control purposes. In fact, solving a probability model has more or less the same implication. Solving a structural equations model like the one used in the American Customer Satisfaction Index computation will yield some indices like customer satisfaction or customer loyalty. Usually, solving a model first requires estimation of model parameters from observed data.

Model-building may also include model specification.

7.3 Types of Models

Models are usually classified as iconic or scale models, analogue models and symbolic models. These apart, there are simulation models and operational gaming models. An iconic model *reproduces* properties or features of interest as possessed by the real-life entity (system or process or phenomenon) on a different – usually a diminished – scale in the model. The model of a building developed and sometimes kept on display even after the building has been completed or a clay model of the physiography of a certain region or a toy passenger bus with toy passengers inside are all examples of scale models. In analogue models, properties of the modelled entity are *represented by some other properties or features of the model*. Thus a map is an analogue model where different ranges of heights or depths are indicated by different colour bands. The flow of electricity is sometimes modelled by the flow of water through pipes. Both the iconic and the analogue models are concrete and specific for each modelled entity (an analogue model may sometimes be used for a class of modelled entities). Hence these models are not amenable to convenient interventions or manipulations to study cause-effect or input-output relations. Even then they are developed and used for specific systems and processes in engineering investigations. The globe is an iconic model of the Earth while a map is an analogue model.

Easy to manipulate and quite generic in nature to have applicability in a whole class of similar systems or processes is a symbolic or mathematical model in which properties of the modelled entity are neither reproduced on a diminished scale nor replaced by some other properties in the model but are represented simply by symbols (numbers, letters and signs). In fact, these models have the widest possible applications in scientific research in different disciplines.

Mathematical models range from a single mathematical equation (algebraic or differential) to an infinite system of probabilistic differential difference equations, from a simple probability distribution to one where a prior distribution is assumed for the parameter(s) in the probability model, from a simple birth process to a complex non-homogeneous stochastic process, from a simple linear regression to a system of linear structural relations, etc.

Any model involves some variable(s) and some relations connecting such variables including some constraints in which certain parameters appear. Such variables could be deterministic or stochastic (random), the relations could be functional or otherwise. Constraints could be stated as equations or inequalities. In the context of optimization models, we can think of three types of variables, viz. decision variables whose values are to be determined to optimize some objective function(s), given variables whose values are given in the problem beyond the control of the decision-maker and consequence variables which define the objective function. Given variables may appear as parameters in the objective function and/ or in the constraints. In a descriptive model, there are no controllable decision variables explicitly stated as distinct from the given variables and the objective function is not to optimize but to describe or predict the state of the underlying system or process in terms of relations connecting the given variables.

Now we can attempt a simple classification of models as deterministic versus stochastic and as descriptive versus prescriptive. Thus a simple growth model given in terms of a simple differential equation with no random variables is a deterministic descriptive model, while the simple inventory problem with known demand for a single period to minimize the total inventory cost yielding the economic order quantity is a deterministic prescriptive model. A waiting line model with arrivals of customers following a Poisson process and random service times with an assumed probability distribution is a stochastic descriptive model. A replacement model for randomly failing items used to minimize the total cost of replacement (considering both cost of planned replacement and cost of replacement on failure) by choosing the time for planned replacement illustrates a stochastic prescriptive model.

Incidentally, all these models are branded as mathematical or symbolic where properties or features of the modelled system or process and relations connecting those are simply indicated by symbols.

Sometimes models used to describe a system can also be used to predict the state of the system at some future time. Thus, a deterministic growth model like the logistic model given by $P(t) = L / [1 + \exp \{r (\beta - t)\}]$ with the three parameters, viz. L (the limiting population size determined by the carrying capacity), β (the point of time when population reaches half of its limiting size) and r (the rate of growth had population grown exponentially) estimated by using past values of $P(t)$, is often used to predict population at a future time t. Similarly the steady-state queueing model assuming constant rate of arrival (birth) λ and constant service completion (death) rate μ indicated by the system of equations

$$(\lambda + \mu)\, p_n = \lambda\, p_{n-1} + \mu\, p_{n+1} \quad n \geq 1 \text{ and } \lambda\, p_0 = \mu\, p_1$$

where $p_n = \lim P_n (t)$ as t tends to ∞ is the steady-state probability that the population size will be n at time t, is also used to predict the state of the system at a future time t.

Models may accommodate constraints on possible parameter values or on their estimates. Choosing one among a class of models may invite different alternative criteria. Estimation of model parameters has continued to attract the interest of statisticians.

We can possibly distinguish between (1) research (by statisticians) to come up with new ideas and methods and also more efficient tools and algorithms to deal more effectively with newly emerging phenomena and associated data types and (2) research to better comprehend some phenomena or systems, characterized by uncertainty.

Usually, some of the existing models is used, based on some evidence or some prior knowledge or even suggested for the sake of simplicity. Often, a test for its validity in the particular context is carried out. On some occasions, robustness of the model solution or model-based analysis against deviations from the assumed model is also examined.

The other and more convincing approach is to develop a model based on certain established and experimentally verifiable features of the system or process or phenomenon. Development of such models require a sound domain knowledge.

A statistician may need to collaborate with a domain expert to identify and incorporate specific features of the modelled entity. This is also required to try out the model developed at any stage on real data and to work out suitable modifications in a bid to provide a better representation of the underlying phenomenon. In fact, in research involving use of probability distributions to represent variations in an observed variable, the data-based choice of a suitable probability model from among the huge kit of available models is the first task in data analysis. Of course, in specific situations, to attempt to develop a model starting from some postulates which can be reasonably taken for granted in the context may be a better alternative.

7.4 Probability Models

7.4.1 Generalities

The phenomenon universally observed is 'variation' across individuals, time and space. And in many cases, such variations are caused by some uncontrollable chance factors over and above some controllable assignable factors affecting the output of a system or the outcome of a process, rendering the latter 'unpredictable'. Thus, variations in weather conditions across days in a crop season are random, as are variations in yield from one plant to another of the same variety receiving the same treatment.

To represent such random variations in some variable(s) as revealed in a sample of observations, we make use of probability models through which we pass from the particular sample to reach some conclusion relating to the phenomenon (population or ensemble) itself. From the existing huge stock of probability models, we can select one that fits our observations best and facilitates inductive inferences. A choice among alternatives can be based on some criterion – usually a penalized likelihood of the sample observations, assuming a particular model as true – which is to be obtained from the sample data. Otherwise, we have to develop a model to incorporate special features of the observed data and of the underlying phenomenon. In fact, many new models have been added to the current kit of models, e.g. exponentiated probability models or skewed normal distribution or generalized exponential or gamma distribution and a host of similar others.

7.4.2 Some Recent Generalizations

While this section is not meant to provide even a brief outline of the whole host of just univariate probability models that have emerged over the years, it may be of some interest to look at some of the more recent attempts to generalize existing probability models to expand the scope of applicability of such models. In fact, new probability distributions are being suggested by way of modifying or extending or generalizing some existing models. This is achieved by working out some transformation of the underlying variable or introducing some new parameters or exponentiating the distribution function or taking some convolution or considering a mixture of a finite or an infinite number of distributions or by considering the relation between the distribution function and some property of the distribution like failure rate for which different forms can be assumed.

Speaking of modifications by introducing a new parameter, the skewed normal distribution introduced by O'Hagan and Leonard (1976) and later studied by Azzalini (1985) is provided by the density function

$$\varphi_1(x) = 2\,\varphi(x)\,\Phi(\alpha x) \qquad \text{for the standardized variate } z = (x - \mu)\,/\,\sigma$$

where α is the skewness parameter, φ and Φ being the ordinate and left-tail area for the standard normal deviate. The mean and variance for the skew normal distribution are respectively mean $= \mu + \delta\,\mu\,(2\,/\,\pi)^{\frac{1}{2}}$ and variance $= \sigma^2\,(1 - 2\,\delta^2\,/\,\pi)$.

The distribution retains some useful properties of the normal distribution. Arnold (2004) provides a characterization of the skew-normal distribution and the generalized Chi (1) distribution. In fact, the squared or the power normal or the exponentiated normal distribution illustrates a modified version of the classic normal distribution. We also have the multivariate form of the skew-normal distribution.

Thus, starting with the exponential distribution with a constant failure, we assume a linear failure rate function as r(x) = a + b x, we get the linear exponential distribution as having the distribution function as F(x) = 1 − exp [− (ax + bx² / 2)]

The Topp-Leone linear exponential distribution will then have the distribution function

$$G(x) = [1 − \exp[− 2\{ax + bx^2 / 2\}]^\alpha \quad \alpha > 0$$

Topp-Leone Rayleigh distribution corresponds to the case when a = 0 and Topp-Leone exponential distribution is obtained by putting b = 0. Bain developed a quadratic hazard rate distribution with distribution function as

$$F(x) = 1 − \exp − (ax + bx^2 / 2 + cx^3 / 3)]$$

Using the same principle of exponentiation (exponent α), we get the extended quadratic hazard rate distribution which will cover the following distributions as particular cases

	a	b	c	α	Name of Distribution
1	a	0	0	α	Extended exponential
2	0	b	0	α	Extended Rayleigh
3	0	0	c	α	Extended Weibull
4	a	b	c	α	Extended quadratic hazard rate
5	a	b	0	α	Linear failure rate

It may be noted that EHQR shows different behaviours of the hazard rate, including bath-tub and inverted bath-tub also for different combinations of parameter values.

Similarly, by first introducing a skewness parameter in a normal distribution and linking it with a beta distribution, we have the beta-skew normal distribution with the density function given by

$$f(x; \lambda, a, b) = 2 / B(a, b) \{\Phi(x: \lambda)\}^{a−1} \{1 − \Phi x; \lambda)\}^{b−1} \varphi(x) \Phi(\lambda x) x \varepsilon R$$

where B (a, b) denotes the beta function and Φ (x; λ) is the distribution function of a skew-normal distribution. In this case the random variable x ∼ BSN (λ, a, b).

With a = b = 1, we get the skew-normal distribution; the beta-normal corresponds to λ = 0. BSN (0, 1, 1) is a standard normal distribution.

We have the Marshall-Olkin (1997) family of distributions obtained by adding a parameter in an existing distribution with survival function S (x) = 1 − F (x) and a probability density function f (x) to derive a distribution with probability density function $g(x) = \Phi f(x) / \left[1 − (1 − \Phi) S(x)\right]^2$ and a failure rate function h′ (x) = h (x) / [1 − (1 − Φ) S (x)] where h (x) is the hazard rate function of the original distribution. A further generalization has been formulated in terms of the survival function S′ (x) = Φ exp [1 − exp (x / σ) ᵝ] / [1 − (1 − Φ) exp {1 − exp (x/ σ) ᵝ] x > 0, Φ, σ and β > 0.

Aly and Behnkerouf (2011) extended the Marshall-Olkin family based on the probability generating function introduced by Harris (1948), resulting in the Harris generalized family of distributions. The survival function of this family is given by

$$S(x) = \{\Phi(S(x))^r / [1 − (1 − \Phi)(S(x))^r]\}^{1/r} r > 0.$$

Combining these ideas, Jose and Paul (2018) introduced the Marshall-Olkin exponential power distribution in terms of the survival function

$$S(x) = \Phi [\exp \{1 − \exp (x / \lambda)^\beta\}] / [1 − (1 − \Phi) \exp\{1 − \exp (x / \lambda)^\beta\}]$$

The Marshall-Olkin generalized distributions can be regarded as compound distributions with the exponential density as the mixing distribution.

Income and wealth are known to follow heavy-tailed distributions with an interior mode. Such distributions are needed to represent the behaviour of errors in several stochastic models, specially in econometrics. The Laplace or Burr or Pearsonean type XII distributions are common choices. The Dagum model, which is also called inverse Burr, is a recent model used in analysis of income data as also in stress-strength reliability analysis. This three-parameter distribution (Dagum, 1996) has the distribution function

$$F(x) = (1 + \lambda x^{-\delta})^{-\beta} \qquad x > 0; \lambda, \beta, \delta > 0$$

with scale parameter λ and shape parameters β and δ.

7.4.3 Discretization of Continuous Distributions

We need models and we have models called discrete probability distributions for discrete random variables which are either countable or categorical (categories correspond to positive integers). Thus, we have the univariate binomial or Poisson or hypergeometric or negative binomial distributions and also the multivariate multinomial distribution. We also note that observations or measurements on continuous random variables as are available to us are essentially discrete. A whole bunch of research papers have reported discrete versions of well-known continuous distributions.

Empirical distributions with power-law tails indicated as below

$$\Pr(X > x) \sim x^{-\alpha}$$

have been commonly observed in a variety of phenomena across many fields like biology, computer science, chemistry, finance, geo-sciences, economics, social sciences etc. A popular continuous distribution with the above heavy-tail property is the Pareto distribution in its various forms.

Several schemes for discretization of continuous distributions have been discussed in the literature. In fact, we can recognize three different discretization schemes through which such versions in respect of many useful univariate continuous distributions have been developed. The one based on the probability density function $f(x)$ for a continuous distribution is to define a probability function for a discrete version of X as follows:

$$f_D(k) = cf(k) \quad k \mu Z \quad \text{where Z is the set of integers and}$$
$$c = (\Sigma f(j))^{-1} \text{ sum over all integers}$$

This scheme has been used to generate the discrete normal (and half-normal) and also the discrete Laplace distributions. The discrete version of the Pareto variable with p.d.f.

$$f(x) = (\alpha - 1) / x^\alpha \ x \geq 1 \text{ yields Zipf's law with probability function}$$

$f_D(k) = 1 / \lambda(\alpha) [1 / k^\alpha] k = 1, 2, 3, \ldots$ where $\lambda(\alpha)$ is the Riemann zeta function defined as $\lambda(\alpha) = \Sigma k^{-\alpha}$, sum from 1 to ∞ and $\alpha > 1.1$.

The second scheme based on distribution function or survival function defines the probability function as

$$f_D(k) = \text{Prob}[k \leq X < k+1] = F(k+1) - F(k) \ k \in Z$$

This approach has resulted in geometric distribution (from the exponential distribution) discrete Weibull distribution and the discrete Lomax distribution in the form

$f_D(k) = [1 / (1 + k /)]^\alpha - [1 / (1 + k /)]^\alpha$ $k \in N$ the set of positive integers.

The third scheme on the positive half of the real line is in terms of the probability generating function (p.g.f.). The p.g.f. of the discrete version is taken as $G(s) = L(1 - s)$ where L is the Laplace transform of X. A host of distributions including the Mittag-Leffler and discrete Linnik distributions along with the geometric and negative binomial distributions have been derived by this approach.

7.4.4 Multivariate Distributions

We now deal almost always with joint variations in a number of variables revealed in what we call multivariate data and we have to speak of multivariate distributions to represent such data. The bivariate normal distribution arose in the context of linear regression, to answer the question:

For which joint distribution of random variables X and Y will (1) the marginal distribution of X be univariate normal N (0,), (2) given $X = x$, the conditional distribution of Y be Normal and (3) the equiprobability contours be similar ellipses centred at (0, 0)?

Multivariate extensions of univariate distributions have been generated in different approaches, given the marginal distributions (identical in form or otherwise) and the interdependence structure or pattern of association among the variables. Generalizing some characterization result for some univariate distribution to the multivariate set-up, we can get multivariate extensions. This has led to multiple forms of the multivariate exponential distribution – some absolutely continuous while others have a non-zero probability for $X = Y$, most retaining the loss of memory property. The constant failure rate property and the loss of memory property of the exponential distribution are pretty well studied. However, extension of any of these two properties in the bivariate set-up is not unique: the failure rate can be defined as $R(x, y) = f(x, y) / S(x, y)$, where $S(x, y) = 1 - F(x, y) = $ Prob $(X > x, Y > y)$ is the survival function (Basu, 1971; Puri and Rubin, 1974). Alternatively, we can take the vector failure (hazard) rate as $R(x, y) = [- \delta \log s(x, y) / \delta x, - \delta \log S(x, y) / \delta y]$ given by Johnson and Kotz (1975) and Marshall (1975). Cox (1972) introduced a conditional failure rate and a failure rate for the absolutely continuous bivariate variable which views the bivariate lifetime as a point process. In this context, we define the total failure rate as a vector

$$\{r(t), r_1(x / y) \text{ for } x > y > 0, r_2(y / x) \text{ for } y > x > 0\}$$

where $r(t) = - d\log S(t, t) / dt$, $r_1(x / y)$ F X, y) / f_x f (x, y) dx. It can be noted that r (t) is the failure rate of min (X, Y) or of the series system composed of component lives X and Y, and $r_1(x / y)$ or $r_2(y / x)$ are the conditional failure rates.

The following bivariate exponential distributions are characterized by a constant total failure rate.

Freund's model (1961) with joint density

$$f(x, y) = \alpha'\beta \exp[-\alpha'x - (\alpha + \beta - \alpha')y] \qquad \text{for } x > y > 0$$
$$= \alpha\beta' \exp[-\beta'y - (\alpha + \beta - \beta')x] \qquad \text{for } y > x > 0$$

Total failure rate = $(\alpha + \beta, \alpha', \beta')$ where $\alpha, \beta, \alpha', \beta' > 0$.

The Marshall-Olkin model (1967) with joint survival function

$$S(x,y) = \exp\left[-\lambda_1 x - \lambda_2 y - \lambda_{12} \max(x,y)\right] \qquad \text{for } x, y > 0$$

$$\text{Total failure rate} = (\lambda_1 + \lambda_2 + \lambda_{12}, \lambda_1 + \lambda_{12}, \lambda_2 + \lambda_{12}) \qquad \lambda_1, \lambda_2, \lambda_{12} > 0$$

The Block-Basu model (1974) with joint survival function as

$$S(x,y) = \left[\lambda / (\lambda_1 + \lambda_2)\right]\left[\exp\left[-\lambda_1 x - \lambda_2 y - \lambda_{12} \max(x,y)\right] - \left[\lambda_{12} / (\lambda_1 + \lambda_2)\right]\right]$$
$$\left[\exp\left(-\lambda \max(x,y)\right)\right] \quad \text{where } \lambda = \lambda_1 + \lambda_2 + \lambda_{12}; x, y > 0$$

$$\text{Total failure rate} = (\lambda, \lambda_1 + \lambda_{12}, \lambda_2 + \lambda_{12})$$

The Proschan-Sullo model (1974) with joint survival function as
 $S(x, y) = \alpha S_A(x, y) + (1 - \alpha) S_S(x, y)$ where the density for the first absolutely continuous part is given by

$$f_A(x,y) = \lambda_2(\lambda_1 + \lambda_{12}) / \alpha \exp\left[-(\lambda_1' + \lambda_{12})x - (\lambda - \lambda_1 - \lambda_{12})y\right] \qquad \text{for } x > y > 0$$
$$= \lambda_1(\lambda_2 + \lambda_{12}) / \alpha \exp\left[-(\lambda_{22}' + \lambda_{12})y - (\lambda - \lambda_2 - \lambda_{12})x\right] \qquad \text{for } y > x > 0$$

and $S_S(x, y) = \exp\left[-\lambda \max(x, y)\right.$ for $x, y > 0$ where $\alpha = (\lambda_1 + \lambda_2) / \lambda$

$$\text{Total failure rate} = (\lambda, \lambda_1 + \lambda_{12}, \lambda_2 + \lambda_{12})$$

The Friday-Patil model (1977) with the joint survival function having the same decomposition as in the previous model with the density function component as

$$f_A(x,y) = \alpha' \beta \exp\left[-\alpha' x - (\alpha + \beta - \alpha')y\right] \qquad \text{for } x > y > 0$$
$$= \alpha \beta' \exp\left[-\beta' y - (\alpha + \beta - \beta')x\right] \qquad \text{for } y > x > 0$$
$$\text{and } S_S(x,y) = \exp\left[-(\alpha + \beta)\max(x,y)\right] \qquad \text{for } x, y > 0 \text{ with } 0 < \alpha < 1.$$

$$\text{Total failure rate}(\alpha + \beta, \alpha', \beta')$$

Under certain conditions, the constant failure rate and loss of memory property are equivalent.
   Farlie, Gumbel and Morgenstern (Farlie 1960) proposed the following form of a bivariate distribution function, starting from given marginal distributions F and G and survival functions S_f and S_g respectively for two continuous non-negative random variables X and Y and an association measure α

$$S(x, y) = S_f(x) S_g(y) [1 + \alpha F(x) G(y)] \mid \alpha \mid \leq 1$$

Here the two marginals need not have the same form and the correlation coefficient between the two variables varies between 0 and 1/4.

Mukherjee and Roy (1986) suggested multivariate extensions of univariate life distributions by considering

$S(x_1, x_2, \dots x_p) = \exp[-\Sigma \{R_i(x_i)^\alpha\}]^\alpha$ where $R_i = -\log S_i$ and α is the dependency parameter. The model does not require absolute continuity for the marginal distributions. This form, of course, implies a constant inter-correlation between any pair of variables. It can be shown that if X_i has an IFR distribution for each i, then X following this multivariate extension has a multivariate IFR distribution. The same is true for the IFRA property also. The ME model gives rise to positive quadrant dependence when $p = 2$ for $\alpha \geq 1$. It can be seen that Gumbel's bivariate exponential and the multivariate Weibull distribution due to Hougaard (1986) and Crowder (1989) can be derived from this ME model.

7.4.5 Use of Copulas

The literature on multivariate distributions has grown considerably for modelling dependence among a set of random variables. In this connection, we define a copula as the joint distribution of random variables $U_1, U_2 \dots U_p$ each of which has a uniform distribution, by the distribution function

$$C(u_1, u_2, \dots u_p) = \Pr(U_1 \leq u_1, U_2 \leq u_2, \dots U_p \leq u_p)$$

Obviously, $C(u_1, u_2, \dots u_p) = u_1 \times u_2 \times \dots \times u_p$ if the random variables U_i are independent. In the case of a perfect relation connecting the variables implying

$$U_1 = U_2 = \dots = U_p \text{ with probability one,}$$

$$C(u_1, u_2, \dots u_p) = \min(u_1, u_2, \dots u_p)$$

It can be easily proved that

for any set of random variables $X_1, X_2, \dots X_p$ with joint cumulative distribution function $F(x_1, x_2, \dots x_p)$ and marginal distribution functions $F_j(x) = \Pr(X_j \leq x)$

there exists a copula such that $F(x_1, x_2, \dots x_p) = C[F_1(x_1), F_2(x_2), \dots F_p(x_p)]$.

If each $F_j(x)$ is continuous, then C is unique.

This theorem (due to Sklar) links the marginal distributions together to form the joint distribution and allows us to model the marginal distributions separately from the dependence structure. If F and C are differentiable, then the copula density as the partial derivative of C with respect to $u_1, u_2, \dots u_p$ works out nicely as the ratio between the joint p.d.f. to what it would become if the variables were independent. This means we can interpret the copula as the adjustment needed to make the product of marginal p.d.f.s into the joint p.d.f.

Of the various classes of copulas, the Archimedean copula with the form $C(u_1, u_2, \dots u_p) = \delta[\delta^{-1}(u_1), \delta^{-1}(u_2), \dots, \delta^{-1}(u_p)]$ with an appropriate generator δ has been widely used. In the bivariate set-up, a copula is said to be Archimedean when it can be expressed in the form $C_\varphi(x, y) = \varphi^{-1}\{\varphi(x) + \varphi(y)\}$ with the convention $\varphi^{-1}(u)$ 0 if $u > \varphi(0)$ and when the generator φ belongs to the class Φ of continuous, convex and non-decreasing functions on [0,1] for which $\varphi(1) = 0$. Archimedean copulas appear in the definition of 'frailty' models for the joint distribution of two survival times depending on the same latent variable. In fact, there exists a correspondence between the dependence properties of Archimedean copulas and the ageing properties of the associated life distributions.

The choice $\delta(t) = [-\log t]^{\alpha}$ results in the Gumbel copula while $\delta(t) = 1/\alpha(t^{-\alpha} - 1)$ yields the Clayton copula. Each of these involves a single parameter α to control the degree of dependence. Unlike linear correlation, a copula is invariant under monotone increasing transformations of the random variables.

An interesting application of copulas is in risk management, where risks correspond to right tails of loss distributions. In a portfolio of investments, large losses are often caused by simultaneous large moves in several components. Thus a consideration of simultaneous extremes becomes important and is facilitated by the copula. The lower tail dependence of X_i and X_j is given by

$$\lambda = \lim \Pr[X_i \le F_i^{-1}(u) \mid X_j \le F_j^{-1}(u)] \text{ as } u \text{ tends to zero}$$

This depends only on the copula and is given by $\lim 1/u\, C_{i,j}(u, u)$ as u tends to zero.

The Gaussian copula corresponding to a multivariate normal distribution as the joint distribution of the original random variables has the odd behaviour of having 0 tail dependence, regardless of the correlation matrix.

7.4.6 Choosing a Probability Model

Given a sample of observations on a random variable, choosing an appropriate probability distribution that takes care of relevant features of the observed value or categories and can be used to make analyses and inferences relating to the underlying population (phenomenon) is an important task for any research worker who has to deal with random variables. This problem is itself a problem in inductive inference involving three steps, viz. selection, fitting and testing goodness of fit. Sometimes, selection requires fitting in some sense. There could exist three different situations in this context, as pointed out by Chatterjee (1968). Firstly, we may have adequate past experience to select a particular probability model as the basis for further studies. Secondly, we may have a small sample and we may be somehow compelled to use some standard model like the univariate normal or the bivariate exponential distribution (Marshall-Olkin form). In the third situation, we may not have grouped data, e.g. when covariates have to be taken account of. In such a case, the likelihoods of different fitted models (maximal likelihoods if MLEs are used to fit the models), given the data, have to be suitably adjusted to make the models comparable. Subsequently, simplicity, stability and capability for extrapolation are found for the competing models to choose one.

Imagine a random sample of observations $y_1, y_2, \ldots y_n$ whose log-likelihood based on a model M with $p(M)$ independent parameters with parameters replaced by their MLEs is denoted by L. Two criteria, viz. the Akaike information criterion (AIC) and the Bayes information criterion (BIC), are defined as $-2L + 2p(M)$ and $-2L + p(M)\log_e n$ respectively.

Given two competing models, one has to choose the one with a lower value for the preferred criterion. While AIC has been suggested on intuitive grounds with a penalty for each parameter in M to be estimated independently, BIC has been justified on Bayesian analysis using an impersonal prior.

Beyond choosing a model M, one has to test its goodness of fit by using, say, the chi-squared goodness-of-fit test applied to the differences between observed frequencies and frequencies given by the filled model.

7.5 Models Based on Differential Equations

7.5.1 Motivation

In some cases, the behaviour of a phenomenon may be, at least to a reasonable degree of approximation, represented by a simple deterministic model which can be subsequently modified to accommodate random variations exhibited by the phenomenon.

It may be incidentally mentioned here that solutions of some differential equations to be satisfied by the probability density function $y = f(x)$ of a random variable X have yielded some well-known families of distributions. Thus, the normal distribution $N(\mu,)$ can be derived as the solution of

$$dy / dx = - y (x - a) / b_0 \text{ with } \mu = a \text{ and } = b_0$$

Generalizing, we may set up the equation as

$$dy / dx = - y (x - a) / [b_0 + b_1 x + b_2 x^2]$$

which has resulted in the Pearsonean system of unimodal distribution with a degree of contact at both the ends.

7.5.2 Fatigue Failure Model

Let us consider a model to represent the phenomenon of fatigue failure and to work out a model for time-to-failure of a metallic structure subject to random stress in successive stress cycles. There can exist two strategies for predicting reliability. A simple strategy would be based directly on time-to-failure or, in this case, number of stress cycles to failure. The other takes into account growth of a crack due to stress as caused by different physical factors.

Whatever model is chosen, our aim will be to specify the length of a crack after N cycles of the load, which we denote by $a(N)$. Given this, the lifetime of a component is easily defined. Since, near failure, the crack grows very quickly, one can specify lifetime as the time at which the length of a crack first exceeds a suitable threshold length A_{th}. Thus the time-to-failure can be represented as

$$Nj = \inf\{N \mid a(N) \geq A_{th}) \quad \ldots \quad (7.5.1)$$

and, under the assumption that $a(N)$ is non-decreasing, the reliability function is

$$\text{Prob}(Nj > N) = \text{Prob}(a(N) \leq A_{th}) \quad \ldots \quad (7.5.2)$$

A deterministic model of fatigue crack growth may be considered to be a mathematical system which will allow one to make accurate predictions about the lifetime of a material or structure, given information about the material properties, details of the geometry, and the actions to which it is subject. A deterministic model would suggest that if one could specify the various parameters exactly, then one would get an exact prediction of the lifetime. Two such models, viz. the Paris-Erdogan equation and the Forman equation, have been quite often used in metallurgical research.

The Paris-Erdogan equation is derived from empirical considerations, and has no real theoretical basis. The equation models the relation between crack velocity and an abstract quantity called the 'stress intensity range', which describes the magnitude of the stress at the crack tip. This range is denoted by ΔK and is usually defined as $\Delta K = Q \, \Delta\sigma \, \sqrt{\alpha}$, where the constant Q reflects the crack geometry and $\Delta\sigma$ is the range in stress per cycle.

The form of the Paris-Erdogan equation is

$$d\alpha \, / \, dN = C \, (\Delta K)^n \quad \text{and} \quad a \, (0) = A_0 \qquad (7.5.3)$$

We note that C and n are regarded as material constants, but that they also depend upon factors such as frequency, temperature and stress ratio. The Paris-Erdogan equation gives good results for long cracks when the material constants are known, but we note that a large effort is required to determine them, since they are functions of many variables.

The Forman equation. The stress ratio $R = \sigma_{min} \, / \, \sigma_{max}$ (ratio of minimum to maximum stress in a cycle) has been shown to have an important effect on crack growth, according to Bannantine et al. (1989). The Forman equation accounts for the stress ratio, and has the following form:

$$da \, / \, dN = C \, (\Delta K)^n \, / \, ((1 - R) \, Kc - \Delta K) \, \ldots \qquad (7.5.4)$$

where Kc is a critical level for the stress intensity, corresponding to unstable fracture.

Starting with either of these models, one can derive some stochastic model to take into account considerable variation in the behaviour of crack growth even with the same material and under the same environment. And we can randomize a deterministic model in several possible ways. Thus, we can

(1) Assume random model parameters θ. Since α is a function of θ, a distribution on crack length is implied. Since Nj is a function of α and therefore of θ, a lifetime distribution is also implied.

(2) Take some non-decreasing stochastic process (such as a birth process) indexed by N and specify process parameters such that the expected value of the process is $\alpha \, (N)$.

(3) If $\alpha \, (N)$ has been defined in terms of a differential equation, one can form an equivalent stochastic differential equation whose solution is a stochastic process model for $\alpha \, (N)$.

Random parameter model. A simple random model for $\alpha \, (N)$ is to take the Paris-Erdogan model and assign probability distributions to the parameters C, n and A 0. Here, we give another example, this time from the theory of ceramic materials (Paluszny and Nicholls).

For ceramics, it is more relevant to assume the material is under a constant stress σ, instead of a cyclic one, in which case one indexes growth by time t and the equivalent to the Paris-Erdogan equation is

$$d\alpha \, / \, dt = v0Kn, \, \ldots \qquad (7.5.5)$$

where $K = Q \, \alpha \, Ö \, \sigma$ is the stress intensity, and v0 and n are material-specific parameters.

The approach to finding the lifetime distribution is different, as it is based on modelling the strength of the specimen rather than crack length. The strength S(t) (or maximum load that the specimen can bear) at time t is related to crack length by S for a material constant B; this equation is derived in Griffith's seminal work. So as a (t) increases, the strength decreases and failure occurs as soon as S (t) < σ:

$$Tj = \inf \{t \,|\, S(t) < \sigma\}....$$
(7.5.7)

Using Equations 7.5.5, 7.5.6 and 7.5.7, and given that n > 10 for ceramic materials, one can show that the lifetime is approximately

$$Tj \approx 2 / \{(n-2) v0 \, Bn \, \sigma n (\sqrt{A}0) n - 2\} \quad ...$$
(7.5.8)

The model is randomized by considering the initial strength S (0) – which is related to the initial crack length by Equation 6 – to be Weibull distributed. The implied distribution of Tj is also Weibull, thus justifying one of the most common reliability models.

7.5.3 Growth Models

Growth (including decay and decline) is the most common phenomenon we come across in many areas of investigation. Growth of a population of living beings including micro-organisms or of inanimate entities like land areas put under different crops requires quantification and explanation. Thus the growth of a human population along with growth in urbanization and industrialization can provide an explanation of growth in the proportion of land area put under non-agricultural purposes. And the consequence of the latter could be linked to changes in agricultural practices meant to increase the productivity of the land. Changes in cropping patterns adopted by the farming community can be linked to increase in gross value added per unit area mandated by a change in the living conditions of the farmers.

Whenever we speak of changes – essentially over time – we deal with dynamic models, as opposed to static ones which are focused on an existing situation only. Most often differential or differential-difference equations – a single equation or a set – are used as models with parameters that may be constants or time-dependent or even random. Thus the logistic law of growth, which has undergone many modifications and extensions, was derived from a simple differential equation

$1 / P(t) \, d \, P(t) / dt = [1 - k \, P(t)]$ the l.h.s. corresponds to the relative rate of growth at time t with P (t) denoting population size at time t and k is a constant that is related to the limiting population size. In fact L = 1 / k is taken as the size of the limiting population. The solution that can be used to predict the population at a future time t is given by

$$P(t) = L / [1 + \exp \{r (\beta - t)\}]$$

Of the three parameters L, r and β it may be noted that L = P (∞) is the limiting population size, linked with concept of 'carrying capacity' of a geographical region in terms of its resources and sinks to absorb wastes to support human and the linked animal population, β is the time at which P (β) = L / 2 and r is the growth rate which would have been observed in a situation where the relative rate of growth remains a constant and does not decline with population size. The logistic curve looks like a laterally stretched S (sigma) and is regarded as the sigmoid or simply the S-curve. Its use now goes much beyond population analysis and covers such areas as technology growth and growth in the number of innovations and the like. The asymptote indicated by L is the limiting population size in a closed community with no migration in or out.

7.6 The ANOVA Model

In many studies the data we eventually collect can be grouped according to some associated characteristic(s) – nominal or categorical or numerical – usually by design, so that differences among the different groups in respect of the variable(s) of interest can be appropriately revealed and, whenever needed, tested for their significance. The totality of units or individuals may be classified or grouped according to k different features, each of which contributes to the total variation among the n units in respect of a single variable of interest, say y. In analysis of variance, we split up the total variation in y (total sum of squares of deviations from the overall mean) into k + 1 components, k corresponding to the k groupings and a remaining one which takes care of the unexplained part, This residual variation is sometimes referred to as 'error', though it is not due to any error in measurements or in grouping etc. It simply implies that a part of the total variation cannot be explained by the groupings and can be ascribed to uncontrolled 'chance' causes of variation.

7.7 Regression Models

7.7.1 General Remarks

In the large category of explanatory models, we have the regression models and structural relations models. Regression models depict dependence relations connecting different phenomena and the associated variables. And these models are often used to make predictions. Essentially meant to relate phenomena taking place concurrently or to relate one current phenomenon with another that happened earlier, regression models are widely used in all scientific studies. We have a large variety of regression models depending on the nature of the variables involved and the nature of relations connecting them. Somewhat different are structural equation models where a variable that plays the role of an independent regressor or explanatory variable in one equation can become the dependent variable in another equation. Usually, these are linear relations that connect some exogenous with some endogenous variables. Some of the variables could be latent, besides the usual manifest variables. Models representing socio-economic phenomena, which are not uncommon in agriculture, belong to this category. Regression models with all the ramifications and applications along with associated problems of estimation and testing of hypotheses occupy the centre stage in modelling and analysis exercises in data science.

In the simplest se-up of the dependent variable y depending on a single non-stochastic independent variable X, we work with both linear and polynomial regressions, estimating the regression coefficients by the method of least squares or the method of maximum likelihood.

A lot remains to be done in estimating regression parameters subject to some constraints imposed on them like in a linear regression involving two independent variables X_1 and X_2 affecting the dependent variable Y in the form

$$Y = \alpha + \beta_1 X_1 + \beta_2 X_2 + e \text{ with e as the random error}$$

it may be natural to require that $\beta_1 > \beta_2$ or the other way round or, simply that $\beta_1 > 0$.

Regression models are subject to deterministic and chance constraints on the regression coefficients like $\beta_i < \beta_j$ or in general $\Sigma l_i \, \beta_i = 0$. We can also come across a situation when any such constraint holds with some probability. In such cases, estimation of the regression coefficients and examination of the distributional properties of the estimates pose some problems. Such constraints in the model are often suggested by the experimenter or some domain expert and the statistician starts working on such problems.

7.7.2 Linear Multiple Regression

The most often used regression model is given by the equation

$$y_k = \beta_0 + x_k' \, \beta + e$$

where $(y, x_k) = (y, x_{1k}, x_{2k}, \ldots x_{pk})$ is the observation for individual k, k = 1, 2, ... n and e is the error (difference between the observed and the fitted linear relationship). In matrix notations, we can write this as

$$y = X \beta + e$$

The following assumptions are generally made.
(1) Errors e are IID random variables, normally distributed with mean 0 and a constant variance, (2) X has full rank p, (3) X is a non-stochastic matrix and (4) $\lim (X \, / \, X \, / \, p)$ exists and is a non-stochastic and non-singular matrix as p tends to infinity.

Regression coefficients β_k are estimated by the method of least squares or by the method of maximum likelihood and we can also derive best linear unbiased estimates. Goodness of fit is usually checked by the coefficient of determination R^2 or adjusting for the degrees of freedom to use $R_c^2 = 1 - (n-1)(1 - R^2) / (n - p - 1)$. Adequacy of regression analysis can be checked by the usual F test considering the SS due to regression and SS due to residual (error). The choice among several candidate models is based on the adjusted coefficient of determination.

If the number of data points n is smaller than the number of explanatory variables, the problem of multi-collinearity comes up in the ordinary least-squares procedure for parameter estimation. Here the method of partial least squares (PLS) converts the original model to $y = \alpha + \beta_1 + {}_2 + \ldots + \beta_p$ to $Y = \beta_0 + \beta_1 \, T_1 + \beta_2 \, T_2 + \ldots T_p$ where T_i values are linear, uncorrelated combinations of the original regressors' X_i values. PLS starts with centring all variables around their respective means and working out simple linear regressions of y on each T_i and generates an estimate of y as the ordinary or weighted mean (with weights as corresponding data values x_i) of the estimates offered by the individual regressions. A regression of X_i on T_1 yields a residual that reflects the information contained in X_1. Similarly, we get residuals from the other regressions. Using the regression coefficients here we can take the average of predictors of y as a second predictor of y, viz. T_2. This iteration continues till we stop at T_p. Eventually the parameters β are estimated by the ordinary least-squares procedure.

7.7.3 Non-Parametric Regression

In non-parametric regression, we do not specify the relation between y and the explanatory variables **X**. This model is written as $y = \varphi(X) + \varepsilon$ where $\varphi(x)$ is the unspecified function of **X** at some fixed value x given by $\varphi(x) = E(y \, / \, X = x)$. The first derivative of φ indicates the regression or response coefficient and the second derivative indicates the curvature of φ. We assume that both y and $X_1, X_2, \ldots X_p$ are stochastic and $E(\varepsilon / X = x) = 0$ and $Var(\varepsilon / X = x) = \sigma^2 I$. The regression coefficient of y on x_j is $\beta_j = \delta \varphi(x) / \delta x_j = \lim [\varphi(x+h) - \varphi(x)] / h$ as h tends to ∞, where $\varphi(x-h) = \varphi(x_1, x_2, \ldots x_{j-} h, \ldots x_p)$. The literature cites several kernel-type estimators besides the weighted least-squares estimator of φ.

If p is large, to capture the main effects of the explanatory variables, we use classification and regression trees using a recursive partitioning algorithm of the covariate space, which results in a tree structure. We have binary splits like $S_1 = \{x: X_j \leq x)\}$ as the left node and $S_2 = \{x: X_j > x\}$, where j is the splitting variable. The idea is to ensure that values of y are as homogeneous as possible within a node and as different as possible from those in the other node. Each of the nodes is again split into two by using the same or a different splitting variable. Thus a tree structure is developed. At the final stage, values of y in any leaf node are averaged. A procedure called GUIDE (generalized, unbiased, interaction detection and estimation) proposed by Loh (2002) and expanded by Kim (2007) has been explained in Rao et al. (2008).

7.7.4 Quantile Regression

Quantile regression provides a tool to directly investigate the upper (lower) conditional quantiles of interest in cases of structural heterogeneity across the conditional distributions. It is quite possible that a researcher is interested in the behaviour of the quantiles of interest rather the conditional means with changes in the independent variables or factors. In fact, when the assumption of constant conditional variance required in least-squares estimation of the usual regression based on conditional means is violated, e.g. when the conditional variance increases with an increase in the independent variable, it may be useful to try out quantile regression. Koenker (2005) mentions a wide array of applications of quantile regression analysis.

In the regression of systolic blood pressure on age and diastolic blood pressure, the upper conditional quantiles of systolic blood pressure are usually of interest and can be differently related compared to the conditional means. In fact, diastolic blood pressure (DBP) is less responsive as people get older than systolic blood pressure (SBP). SBP alone is elevated while DBP remains within normal range. Here we would like to study the constrained regression where the coefficient of DBP is required to be a non-decreasing function of age to correctly reflect this progressive relationship in response to ageing. Quantile regression with shape-constrained varying coefficients has been studied by Kim (2007) using splines.

For a random vector (U, X, Y) the rth quantile of Y given (U, X) = (u, x) is denoted as $q_r(Y | u, x)$. Suppose this is linear in x with unknown smooth functions of u as the coefficients in the form

$$q_r(Y | u, x) = \beta_0(u) x_0 + \beta_1(u) X_1 + \ldots + \beta_p(u) x_p$$

where the shape information for some $\beta_j(u)$ are available a priori and the estimators are required to satisfy the shape constraints. Without shape constraints, Honda (2004) and Kim (2007) provide estimates of coefficients in the case of IID data and Cai and Xu (2002) and also Xu (2005) provide results for time-series data. With constraints in the least-squares framework, the problem has been studied by Orbe and Rodriguez (2005) and Kim (2007).

7.7.5 Artificial Neural Network (ANN) Models

Somewhat similar to regression models in terms of objectives and having a wider scope, artificial neural networks are inter-connected networks of neurons which are elementary (information) processing or computational units. Unsupervised network models are used for pattern recognition and pattern classification. As mentioned in Wikipedia, an ANN is based on a collection of connected units or nodes called artificial neurons, which loosely model the neurons in a biological brain. Each connection, like the synapses in a human or animal brain, can transmit a signal from one artificial neuron to another. An artificial neuron that receives a signal can process it and then signal additional artificial neurons connected to it. A single-layer perceptron model consists of several neurons forwarding their outputs to another neuron that combines the inputs with some weights and generates an output through an activation or response function. A multi-layer perceptron type ANN consists of one input layer, one or more hidden layers and one output layer – each layer with several neurons and each neuron in a layer connected to neurons in the adjacent layer. Except for the input layer, each neuron receives signals from the previous layer, linearly weighted by the interconnection values between neurons. The neuron produces its output signal usually through a sigmoid function.

A single-unit perceptron is a simple illustration of a feed-forward neural network, with k input units $x_1, x_2, \ldots x_k$ processed by a single neuron to produce the output y (expected response) as $Y = w_0 + \Sigma\, w_i\, x_i$ or more generally as $y = f(x; w)$ f being the activation function. Given a training sample $[x^{(i)}, y^{(i)}\; I = 1, 2, \ldots n]$ on n individuals, weights w are determined in such a way that the energy or learning function $E(w) = \Sigma\, \{y^{(i)} - f(x^{(i)}, w\}^2$ is minimized. In practice several numerical methods are used to solve this least-squares problem, the most well-known being the generalized delta rule or the error back propagation.

A useful application of ANN for short-term prediction of demand for electric power has been discussed by Park et al. (1991). Traditional computationally economic approaches like regression and interpolation may not give sufficiently accurate results. In such models, the current load is related functionally to weather variables and with short-term weather forecasts load in the immediate future is predicted. The authors use an ANN approach to load forecasting. Past data help in building up the non-linear relation between load and weather variables, without assuming any functional form. A generalized delta rule is used to train a layered perceptron type ANN, which is trained to recognize the distinct peak load of the day, total load of the day and hourly load. There have been many other interesting and useful applications of ANN in diverse fields. In the field of medicine ANN with a logistics regression model for the activation function has also been reported in the literature.

7.8 Structural Equation Modelling

Models to represent structural relations connecting elements and their features within a system were developed through interactions among social scientists (including economists), geneticists and statisticians. And many problems in choosing and fitting these models and even in interpreting the results of analysis based on such models attract statisticians. Structural equation modelling (SEM) is a statistical exercise for testing and estimating *causal relations* using a combination of *statistical data* (observed as well as unobservable) and *qualitative causal assumptions*. SEM allows both exploratory and confirmatory analysis, suiting theory development and also theory testing and validation. Path analysis, factor analysis and linear regression analysis are all special cases of SEM.

Structural equation models appear to be multivariate linear regression models. However, unlike in the latter, the response variable (to be predicted/estimated by the model) in one (regression) equation in SEM may appear as a predictor in another equation. Variables may influence one another reciprocally, directly or through other variables as intermediaries.

Two kinds of variables appear in SEM. Exogenous variables are external inputs to the system which never depend (regress) on other variables within the system, although other variables within the system depend on such variables. Endogenous variables depend (regress) on other variables and are produced within the system.

In a directed graph representing such a model, arrows can only emanate from exogenous variables, while arrows can emanate from and terminate in endogenous variables.

- Variables could be directly measured /observed or could be unobservable.
- The latter are referred to as latent variables related to *traits* or *factors*.

We have to develop good indicators for the above, maybe in terms of responses to some questions or statements (loaded with the factors or traits) suitably scaled.

Two main components of an SEM are the *structural* model showing potential causes of dependencies between exogenous and endogenous variables and the *measurement* model showing the relations between latent variables and their indicators (empirical variables). Factor analysis – exploratory or confirmatory – contains only the measurement model, while path analysis contains only the structural part.

Pathways can posit two types of relations:

(1) free pathways in which hypothesized (in fact, counterfactual) relations between variables are tested and are left 'free' to vary and
(2) relationships between variables that have been already estimated, usually based on previous studies, and are 'fixed' in the model.

Models in SEM can be of different types.

The observed-variables model considers relations (taken to be linear) among manifest variables only, sometimes with appropriate lags in time. Klein's macro-economic model representing the US economy over a period of time is an oft-quoted illustration of such a model. Parameter estimation in such a model is relatively simple.

A recursive model, on the other hand, generally involves both manifest and latent variables and is often shown as a graph (path diagram) following the conventions:

directly observable variables in squares

unobservable variables in circles/ellipses

exogenous variables as x's,

endogenous variables as y's and

disturbances as ε's

single-headed arrows are structural parameters

double-headed arrows represent non-causal, potentially non-zero, co-variances between exogenous variables (also between disturbances)

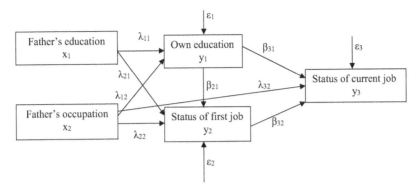

FIGURE 7.1
Status of current job linked to endogenous and exogenous variables.

Structural parameters (coefficients) relating an endogenous variable to an exogenous variable are shown as λ's and those relating one endogenous variable to another are shown as β's.

An illustration can be seen in Figure 7.1.

The structural equations in this model can be read off the path diagram as

$$Y_{1i} = \lambda_{10} + \lambda_{11} x_{1i} + \lambda_{12} x_{2i} + \varepsilon_{1i}$$

$$Y_{2i} = \lambda_{20} + \lambda_{21} x_{1i} + \lambda_{22} x_{2i} + \beta_{21} y_{1i} + \varepsilon_{2i}$$

$$Y_{3i} = \lambda_{30} + \lambda_{32} x_{2i} + \beta_{31} y_{1i} + \beta_{32} y_{2i} + \varepsilon_{3i}$$

In recursive models, there are no reciprocal directed paths or feedback loops and different disturbances are independent of one another (and hence are unlinked by bi-directional arrows). Hence, the predictors in the structural equations are always independent of the error in that equation and the equation can be estimated by OLS regression. Estimating the above model is simply a sequence of OLS regressions.

LISREL models are more general. The path diagram (Figure 7.2) represents peer influences on the aspirations of high-school male students. Two of the endogenous variables, viz. Respondent's General Aspiration (μ_1) and Friend's General Aspiration (μ_2), are unobserved variables. Each of these has two observed indicators: the occupational and educational aspirations of each boy – y_1 and y_2 for the respondent and y_3 and y_4 for his best friend.

The following issues should attract the attention of any research worker using SEM.

- Specification – specially when SEM is used as a confirmatory technique.
- Sometimes a set of theoretically plausible models may be specified to assess whether the model proposed is the best of the set.
- Estimation of free parameters (sample size etc.)
- Assessment of model and model fit
- Model modification, when needed and
- Interpretation and communication

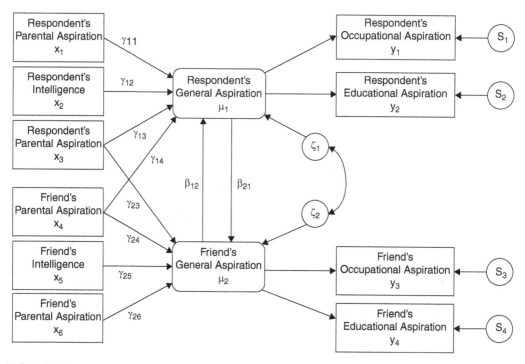

FIGURE 7.2
Poor influences on aspirations.

7.9 Stochastic Process Models

Stochastic process models are involved whenever we are interested in a time-dependent phenomenon or a phenomenon that is affected by contagion or proximity in space. In fact, in such situations any variable observed or measured at different times or at different locations are no longer independent as in the case of models where we deal with independently (and most often identically) distributed random variables. Patterns of dependence of current observations on previous ones are modelled by transition probability matrices and these probabilities themselves could be time-dependent or time-variant. The system being studied can be in different (finitely many or infinitely many) states and one can observe the system at discrete points of time or continually over time. The simple Markov chain and the Poisson process are widely used in modelling a wide range of systems in process control, market research, reliability analysis, social and occupational mobility studies, staffing problems, and many others.

A simple and useful application of a finite Markov chain which always admits of a limiting stationary distribution is given by an attribute inspection classifying successive items produced as either defective or non-defective. In the case of a continuous (as against a batch-by batch) production process where some disturbance in the process will affect some consecutive items in respect of quality due to contamination, the following transition probability matrix P provides the basis for any control scheme.

		Next Item	
		defective	Non-defective
Current	Defective	p_{11}	p_{12}
Item	Non-defective	p_{21}	p_{22}

The steady-state distribution is given by π_1 = proportion of defective items in the long run and $\pi_2 = 1 - \pi_1$ and π_1 can be used to develop control limits on a control chart. These proportions (probabilities) are obtained from P as $[p_{21} / (p_{12} + p_{21}), p_{12} / (p_{12} + p_{21})]$.

The mover-stayer model has found many useful applications in market research and also in studies on social mobility across generations and occupational mobility across generations and also within a generation. Let $p_{ij}^{(t)}$ be the probability of transition from the ith class at time (in generation) t to the jth class at time (in generation) t + 1. Let there be k classes, social or economic or occupational. The $\Sigma_j p_{ij}^{(t)} = 1$. Also let $\pi_i^{(t)}$ be the proportion of the population under consideration at time t belonging to class i. Then the dynamics of the situation as revealed over successive time periods / generations is indicated by the equation $\pi^{(t+1)} = \mathbf{P} \, \pi^{(t)}$, where P is the transition probability matrix and $\pi^{(t)}$ gives the population distribution across the k classes at time (in generation) t. Assuming the steady-state transition probability matrix to be $\mathbf{P}$ independent of time, we can write $\pi^{(t)} = \mathbf{P}^t \, \pi^{(0)}$, where $\pi^{(0)}$ denotes the initial distribution.

In the above set-up, Σp_{ii} is the proportion of individuals who stay in the same class in both the times (generations) while the sum of the remaining elements in $\mathbf{P}$ defines the proportion of movers. Mobility patterns corresponding to perfect mobility, perfect immobility and extreme movement are characterized by different properties of $\mathbf{P}$. In fact, perfect mobility corresponds to a matrix with identical rows $[r_1, r_2, \ldots r_k]$ with $\Sigma r_i = 1$. In particular, one can take $r_i = 1/k$ for all I. If $\mathbf{P}$ is a diagonal matrix, it characterizes perfect immobility with no 'movers'. A matrix with all diagonal elements zero or with no 'stayers' reveals extreme movement. Different measures of mobility can be defined in terms of the matrix $\mathbf{P}$ along with the initial distribution or in terms of divergence between two successive distributions. To study occupational mobility, one can have recourse to renewal reward theory, reward corresponding to increase in income or other benefit associated with a transition from one occupation to another.

The reliability of a system has often been studied in terms of the damage done to the system due to shocks received by the system over time (during its operational life). Shocks arrive at random points of time according to some stochastic process and cause random amounts of damage. Once the cumulative damage exceeds a certain limit, the system fails. This generates the failure-time distribution and hence provides the expression for reliability or probability of survival till time t.

Replacement of failed units or units going to fail shortly is guided by renewal theory.

Just to illustrate the use of stochastic models in representing the outputs of production systems, we consider a model that has been applied to study the performance of a power-generating system. The Balleriaux model for this purpose makes the following assumptions:

1. The system consists of n generating units which are brought into operation according to some merit order of loading. The simplest order is the one that is based on operating costs. That is, the cheapest unit is dispatched first, followed by the next cheapest unit, etc., until the load is met.

2. The ith unit in the loading order has capacity c_i (MW), variable operating cost $d_i(\$/$ MWH) and steady-state unavailability p_i. the unavailability measure is usually referred to in the power system literature as the forced outage rate (FOR). This parameter is estimated from data on the proportion of time for which the unit is unavailable due to unscheduled outages. Units are assumed to be available/unavailable independently.

3. The load values are known for each hour during the study interval [O,T]. Denote the empirical distribution function for this data set by $F(u)$, and define $\bar{F}(u) = 1 - F(u)$. The function $\bar{F}(u)$ is known in the power system literature as the load duration curve (LDC). The LDC gives, for any load level u, the proportion of time during the study horizon that the hourly load exceeds u. Finally, define a random variable U with the property that its survivor function is given by $\bar{F}(u)$.

Define X_j to be the (steady-state) unavailable capacity for unit j at any given time during the interval [O,T]. Then, X_j has the following probability distribution:

$$X_j = \begin{cases} c_j \text{ with probability } & p_j \\ 0 \text{ with probability } & 1\text{-}p_j \end{cases} (j=1,2,\ldots,n)$$

Let $Z_j(T)$ denote the energy produced by unit j during the time interval [O,T]. Then, according to the Balleriaux formula, the expected value of the energy produced by this unit during the interval [O,T] is

$$E[Z_j(T)] = T(1-p_j)\int_{c_{j-1}}^{c_j} \Pr\{U + X_1 + X_2 + \ldots + X_{j-1} > x\}dx$$

where $C_k = c_1 + c_2 + \ldots c_k$ k = 1, 2, ..., n and $c_0 = 0$.
The expected value of the production cost for the system during this interval of time (denoted by K(T) is given by

$$E[K(T)] = \sum_{i=1}^{n} d_i E[Z_i(T)]$$

The important thing to notice here is that these formulae depend only on the distributions of the random variables U and the X_j values. That is, if the steady-state unavailability of the generating units is known, it is possible in principle to compute the expected production costs.

4. The operating state of each generating unit j follows a two-state continuous-time Markov chain $Y_j(t)$ which is in steady state:

$$X_j(t) = \begin{cases} 1 \text{ if unit j is up at time t} \\ 0 \text{ if unit j is down at time t} \end{cases}$$

Let λ_j be the failure rate for unit j and μ_j be the repair rate for unit j. The steady-state unavailability and these transition rates must satisfy the relation

$$\frac{\lambda_j}{\lambda_j + \mu_j} = p_j$$

This ensures that the transition rates are indeed consistent with the unavailability measure.

5. For $i \neq j$, $Y_j(t)$ and $Y_j(s)$ are independent for all t and s.

6. The up and down process of a generating unit continues whether or not it is in use.

7. The chronological load, u(t), at any time t, is assumed to be a deterministic function of t, and is fully specified. (Again, it needs to be consistent with the survivor function of U, i.e. the LDC.)

7.10 Glimpses of Some Other Models

A large number of models which apply in some specific situations can provide research workers with an insight into the development of situation-specific models. We attempt in the following brief descriptions of two such models, indicating areas of their applications, just for the sake of illustration. The reader may refer to books or journal articles on any model that may be of interest to him (her).

State-Space Models: state-space models are being increasingly used to study macro-economic and financial problems and also problems in neuroscience, control engineering, computer systems, statistical learning and in similar contexts. Such models are useful in the study of a deterministic or stochastic dynamical system that is observed or measured through a stochastic process. These models are almost the same as hidden Markov models or latent process models. A state-space representation consists of two equations, a measurement equation which links the observed variables to the unobserved (latent) state variables, and a transition equation describing the dynamics of the state variables. The Kalman filter, which provides a recursive way to estimate the unobserved component based on the observed variables, is a useful tool for analysis using such models.

The model arose in the context of tracking a spacecraft with location x_t and the data y_t reflect information that can be observed from a tracking device such as velocity and azimuth. In general it can be regarded as a system of n first-order differential or difference equations in place of a single differential or difference equation of order n to represent a dynamical system. It is characterized by two principles. First, there is a hidden or latent state process assumed to be Markovian. The second condition is that the observations y_t are independent, given the states x_t. We do not observe the state vector x_t directly, but only a linear transformed version of it with noise added. Linear Gaussian state-space models have found useful applications in diverse fields.

Proportional Hazards Model: this model is attributed to the family of life distributions for which the failure rate at time t h (t; Z) has the form H (t; Z) = $h_0(t)$ φ (Z; β), where Z = (z_1, z_2, ... z_p). I is the vector of covariates or explanatory variables including stresses imposed on the living being or the equipment; h_0 (t) is a base-line failure rate, φ is a function of Z and $\beta = (\beta_1, \beta_2, \dots \beta_p)$ is the vector of regression coefficients $h_0(t)$, and β may depend on some other parameters also. The following two forms of φ are widely used in practice:

(a) log-linear: $\varphi (Z; \beta) = \exp (Z \beta)$
(b) logistic $= \log [1 + \exp (Z \beta)$

This model and the associated estimation theory proposed by Cox (1972) allows processing of failure-time data without assuming any specified functional form of the failure rate function.

Consider a sample of n items put on life test with observed failure times $t_{(i)}$, I = 1,2, ... k < n, the remaining n − k failure times being censored. $Z_i = (z_{i1}, z_{i2}, \dots, z_{iq})$ is a set of covariates associated with the ith failure time. The key role is played by the conditional likelihood function. Tsiatis (1981) considered the log-linear proportional hazard model and assumed that the failure times and censoring times are conditionally independent of the covariate, assumed to be a random variable. He showed that maximum likelihood estimators based on conditional likelihood are strongly consistent and asymptotically normal.

It can be assumed that components of the covariate vector are functions of time. Dependent covariates can also be incorporated in this model.

7.11 Optimization Models

Optimization models are involved in methods and algorithms for maximization or minimization of different types of objective functions, usually under different constraints, as well as in the context of different operations to be carried out in an optimum manner. These exists a wide array of optimization models to represent complex problems or operations which do not admit of optimal solutions in closed form. Some of these problems pose difficulties even in obtaining numerical solutions. In situations where we deal with integer decision variables or some approximate relations defining the constraints or a non-regular objective function and it becomes almost impossible to locate the optimal solution, we tend to find a near-optimal solution or a weakly optimal solution.

The general formulation of an optimization problem can be stated as:

To maximize (minimize) a scalar objective function $Z = f (X; Y_1)$, where $X \in S$ (the set of feasible solutions or vectors which satisfy some constraints, linear or non-linear, involving X and some other given variables Y_2). It is possible that a set of objective functions f need to be optimized simultaneously. Symbols used in this formulation are interpreted as follows.

X represents the vector of controllable decision variables and represents the strategy. We need to determine the values of these variables.

Y_1 denotes a set of given variable values which appear as coefficients in the objective function and yields the value of the objective function for a given strategy. This is often referred to as the state of nature in the context of decision analysis.

Y_2 represents the set of given constants which appear in the constraints on X. In any problem, Y_2 includes one sub-set of given constants corresponding to what are generically called technological coefficients and a second sub-set providing levels of available resources.

The easiest way to illustrate this model would be the linear programming problem which can be stated as below:

Maximize (minimize) $Z = C' X$ subject to $AX \leq (\geq) b, X \geq 0$

We can now recognize C as Y_1 and Y_2 as (A, b). In this formulation, all the variables in X, Y_1 and Y_2 are non-random.

Stochastic optimization has been largely concerned with probabilistic or stochastic models used in operational research, or in decision-making under risk, where we come across randomness in

- the objective function(s) caused by randomness in the 'state of nature' and/or
- the constraints in terms of random given variables or given relations and/or
- the decision variable(s) – by choice.

In solving such models (or the corresponding problems), we deal with probability distributions of the random variable(s) involved – usually assumed to be completely specified – and apply some optimization techniques to suitable deterministic equivalents of the problems, replacing the distributions by some summary measure(s).

In the first view of stochastic optimization, the search for the optimal solution or an approximation thereof does not involve any randomness. And the optimality criterion is not unique, in the absence of a unique summarization of a probability distribution.

Not to be restricted with a completely known distribution, we can try Bayesian analysis, which treats some distribution parameter(s) as random, with some assumed prior distribution(s). Hyper-parameters are assumed as known.

Bayesian analysis can further take into account a sample of observations available on the random variable(s) involved and can proceed on the basis of the posterior distributions. This is particularly suited to situations where only the objective function is random, owing to a random 'state of nature' (in terms of random coefficients in the objective function.

Several decision problems like the secretary selection problem or the problem of locating the optimal design point in terms of the estimated response function in a multi-factor experiment or the problem of determining the parameters in a single sampling inspection plan by attributes are not usually recognized as OR problems. These, however, involve stochastic optimization in a broad sense. Similar is the case with some problems in control theory. In fact, many problems in statistics and its applications also have recourse to stochastic optimization.

A second view of stochastic optimization emerged in the context of various search procedures or meta-heuristics to arrive at some optimal or near-optimal solution to complex optimization problems faced quite often in the context of combinatorial optimization of functions – not necessarily complex or probabilistic – subject to constraints – which also are not always complicated or probabilistic.

Usually, a huge number of possible solutions (not all feasible) exist, and stochasticity in such a context relates to the concrete use of randomness in the search for an optimal solution. Combinatorial optimization may also have to reckon with randomness in the objective function, e.g. in a situation involving random processing times in a sequencing problem.

In the second view of stochastic optimization, we introduce certain probabilities for generating new solutions and/or for accepting or rejecting a new solution. Certain decisions about these probabilities (including the ways these are to be found numerically) have to be taken proactively before the search. There are varying suggestions here.

Three important search methods which can be viewed as stochastic optimization methods are simulated annealing, genetic algorithms and tabu search. Apart from many modifications and ramifications of each of these methods, attempts have also been made to combine interesting and useful properties of the three different methods to develop new methods.

In the first view, the 'stochastic' or 'probabilistic' nature of the objective function and/ or the constraint(s) is emphasized and taken care of in various ways to treat probability distributions and their parameters.

On the other hand, the second view is more concerned with the task of working out the optimal or near-optimal solution to computationally complex problems not involving any randomness in the objective function or the constraint(s).

The first view has a wider ambit of applications and offers a greater scope for theoretical research. The second has enforced the need for 'satisficing' instead of 'optimizing' solutions to many real-life problems which evade easy or closed solutions. This has applicability mostly in discrete search situations and has evoked a lot of studies on computability and computational complexity.

Let us illustrate the first type of stochastic optimization in terms of some exercises in probabilistic or stochastic programming, and then turn over to some decision problems which involve stochastic optimization.

It will be wise to start with the simple linear programming (resource allocation) problem, viz. to minimize the objective function

$$z = C' X \text{ subject to } AX \geq b, X \geq 0.$$

Three different situations can be thought of.

Only C is random, inducing randomness in z which will have a distribution induced by the (multivariate) distribution assumed for the vector C, for each feasible solution X.

Only A or b is random or both these are random, implying chance constraints which are to be satisfied with some specified probability.

Only X is assumed to be random, not as a part of the original formulation of the problem, but to yield a better expected value for the objective function.

There could be complex situations involving randomness in all these three respects.

If we assume C to have a multivariate normal distribution, for a given solution X, z will be univariate normal with $E(z) = \mu' X$ where μ is the mean vector of the coefficients C. Further $Var(z) = \Sigma' X \Sigma$. We take μ and Σ to be known constants.

We can have the following equivalent deterministic problems which can be solved by appropriate programming techniques.

1. Minimize $E(z)$ – linear programming
2. Minimize $Var(z)$ – quadratic programming
3. Minimize the variance (SD) penalized mean $E(z) - k [Var(z)]^{1/2}$ for some given value of k – quadratic programming
4. Minimize the coefficient of variation of z $\{SD(z) / E(z)\}$ – fractional programming
5. Minimize the probability of exceedance $Pr\{z > r\}$ for given $r > 0$ – fractional programming
6. Minimize $z0$ the quantile of z such that $Pr\{z > z0\} = \alpha$ given a small value – fractional programming

The same formulations remain valid even when C has some distribution other than the multi-normal. Obviously, solving such deterministic optimization problems will be quite difficult, specially when the univariate distribution of z is not easily derived and/or not conveniently invertible.

One can imagine similar formulations of optimization problems involving quadratic or polynomial or even posynomial objective functions, though such formulations have not been studied so far.

In all the above formulations, we try to reduce the probability distribution of the random objective function to a single summary measure or a combination of two such measures.

Theoretically, it seems better to establish the dominance of one distribution (indexed by one solution X) over other distributions, in which case the index of the dominating (or dominated) solution becomes the optimal solution. And it is well known that stochastic dominance implies expectation dominance, but is not implied by the latter. Denote the distribution function of z for a given X by F x (t). Then, if for some X/, F X/ (t) ≥ F X (t) for all t > 0, then X/ should be taken as the optimal solution. The corresponding value of z is the stochastically smallest value of z

It is quite difficult, however, to find such an X/ even in simple situations.

In the case of a chance constrained programming problem, we may face several different situations like the following:

1. Minimize z subject to Pr{Ai X ≥ bi} = αi

i = 1, 2, ... m Or Pr{AiX ≥ bi, i = 1, 2, ... m} = α
where α and αi are pre-specified small values (aspiration levels), Ai and bi are the ith row of A and the ith element of b respectively and only b is a random vector.

Assuming the elements of b to be independently (and most often) identically distributed, and that these distributions, say Fi values, are invertible explicitly, these constraints can be re-stated as Ai X= Hi (αi), where Hi is the inverse of Fi I = 1, 2, ... m and a similar relation holds for the overall aspiration level. If we have a common c.d.f. F for all bi, H = F – 1.

2. In the same constraints, elements of A only could be random, not those of b.

The treatment of the chance constraint poses similar problems, as in case 1.

3. We can have randomness in both A and b. To construct deterministic equivalents of the chance constraints may be quite involved, except when the distributions assumed are too simple.

Chance-constrained programming with multiple deterministic objectives can be tackled in terms of goal programming or fuzzy programming. Of course, probability distributions of available resources should be amenable to inversion in closed form.

Generally speaking, there are two broad approaches to solving stochastic linear programming problems, viz.:

- Here-and-now (active) – in which the decision-maker has to make a decision based on the assumed probability distributions and no data
- Wait-and-see (passive) – in which we wait to use some realized values of the random variables involved. In this context, we can simulate such values and reduce the problem to a deterministic optimization problem.

Let us look at the following problem to minimize C' X subject to Pr{AX ≤} ≥ α, X ≥ 0.

This chance-constrained stochastic linear programming problem (SCLP) can be re-formulated so as to minimize f such that Pr{f ≥ C' X} ≥ β subject to Pr{A X ≤ b} ≥ α, X ≥ 0. This really means that f (called the target function) is the quantile of order β in the distribution of z = C' X.

This has been solved through simulation and a theoretical basis has been worked out by Ding and Wang (2012).

The random matrix A and the random vector b are taken to follow a joint distribution Ψ (A, b) and the random vector C has a distribution θ (C) The algorithm proceeds in three steps.

(a) Chance constraint judging algorithm, where we take a sample vector from Ψ, calculate AX, check if AX ≥ b or not. Find this N times and if N' times we find that b ≥ AX and if N' / N > α then we return FEASIBLE, Otherwise, we return INFEASIBLE.

(b) Minimum target function searching algorithm, which goes as follows:

1. Draw N random samples{Ck, k = 1, 2, ... N} from the distribution θ (C).
2. Compute C'k X, k = 1, 2, ... N and arrange the values in ascending order.
3. Set N' as {β N}.
4. Return the Nth largest element in the set {Ck', k = 1, 2, ... N} as the estimate of f.

Ding and Wang also provide a check for the estimation number.

A genetic algorithm based on stochastic simulation has been put forward by Iawamura and Liu (1998) for stochastic chance-constrained linear programming problems.

In the two-stage stochastic optimization framework, decision variables are divided into two sub-sets: (1) variables which must be determined now, before the realization of the random variables (in the chance constraints) is known, and (2) recourse variables which are determined after finding the realized values. Such a problem with simple recourse can be formulated as

$$\text{maximise } \bar{z} = \Sigma c_j x_j - E(\Sigma q_i |y_i||)$$
$$\text{Subject to } y_i = b_i - \Sigma a_{ij} x_j \quad i = 1, 2, ...m1$$
$$\Sigma d_{ij} x_j \le b_{m1+i} \quad i = 1, 2, ...m2$$

$x_j \ge 0, j = 1, 2, ... n$ and $y_i \ge 0$ $i = 1, 2, ... m1$. Here x and y values relate to the first-stage and second-stage decision variables. Values of q define the penalty cost associated with discrepancy between the two sides of the constraint.

Some authors have tried to formulate and solve such problems with interval discrete random variables.

In a somewhat theoretical framework, a few attempts have been made to allow the decision variable(s) to be random and to derive the optimal distribution(s) of such variable(s) that will ensure a better value for the objective function (mostly the expected value of a random objective function). This problem is quite complex even with simple discrete and partly known distributions assumed for the decision variable(s). Most such attempts have dealt with a single decision variable or at most two.

To illustrate the application of stochastic optimization – not, of course, recognized as such – for solving several decision problems, we consider four innocent-looking problems, viz.:

the secretary selection problem, the problem of estimation of parameters in a two-parameter Weibull distribution

the newspaper boy problem and

the single sampling plan for inspection.

While the third is the simplest problem in inventory management within the framework of operational research, the other three do not lie in the usual ambit of operational research. The first one has been analysed as a stopping-rule problem in sequential analysis, the third one is a widely used problem in statistical quality control.

Let us consider the secretary selection problem, which has been examined by various authors. The simplest formulation runs as follows.

There are n candidates who can be ranked according to some selection criterion (criteria) from 1 (best) to n (worst) and there is no tie. We interview I randomly selected candidates one by one and select the best of these I candidates (or the candidate with the smallest rank in the sample). Let r be the rank of the selected candidate in the population of candidates with the expected value E (r).

The problem now is to find i such that E (r) is a minimum. We can make the formulation more acceptable by adding a chance constraint on i such that

$$\Pr\{l\, r - r_0\, l \geq m\} \leq \alpha \text{ (pre-assigned small)}$$

where r_0 and m are very small integers. Let the cost of interviewing a candidate be k, with an overhead cost C. The cost of failing to select the best (among all n) is taken as m (r – 1), m a constant. Thus the total cost to be minimized with respect to i is

$$T = C + k\, i + m\, E\, (r)$$

This innocent problem in the integer variable i is a chance-constrained non-linear integer programming problem. A simple algorithm has been offered by Mukherjee and Mandal (1994). It can be argued that the penalty cost in failing to select the candidate is monotonic increasing in r or E (r), instead of being a constant.

The simplest, and one of the earliest, sampling plan for inspection by attributes which is widely practised may be stated as:

Inspect n items randomly from a lot of N items and count the number x of defectives. If $x \geq c$ (acceptance number), reject and inspect the entire lot; otherwise accept the lot. The problem is to find integers (including zero) n and c such that the amount of inspection is minimized, subject to the requirement that Pr{lot of quality p_t is accepted} $\approx \beta$, often taken as 0.10, p_t being the highest tolerable quality (fraction defective) and the expected amount of inspection from lots of acceptable quality p0 given by $I(p0; n.\, c) = n + (N - n)\, \Pr\{x > c\}$ is a minimum. This is a non-linear integer programming problem with a fuzzy chance constraint, though with a non-random and non-linear objective function.

In the block replacement policy for randomly failing equipment, we generally consider the expected total (failure and preventive replacement together) cost per unit time as a non-random function of time to yield the optimum time for replacement. It will be better to work out the distribution of the total replacement cost and to offer the mode-minimizing solution as optimal. For a risk-averter, it has been shown that this solution is better for 2-Erlang failure-time distribution. In fact, the same advantage can be proved for any IFR (increasing failure rate) distribution.

Similar exercises have been attempted on the cost of maintaining a control chart and acting on its findings by working out the distribution of the cost per unit time

7.12 Simulation – Models and Solutions

Simulation involves the development and deployment of models to artificially resemble the functions or operations of a real-life system or process or situation, usually as the former evolves over time and space. The objective of simulation is eventually to result in models which can represent such systems or situations for the purpose of examining cause-effect relations and can even provide some satisfactory solution to some problem(s) associated with the modelled entity (system or situation). Simulation starts with a model and involves its manipulation so that the simulation model can provide a motion picture of reality. In fact, simulation as a scientific exercise can well be regarded as a replacement for experimentation, which is highly desirable when actual experimentation with the underlying system or situation is costly or time-consuming or hazardous and sometimes impossible.

To solve models representing complex systems, we need some inputs like the behaviours of incoming units to a service system or the dynamics of a physical system which can fail to function if the stress or load exceed a certain limit or just the demand of electric power for industrial consumption during a certain period etc. characterized in terms of some parameters. And to generate such inputs at least provisionally we need to simulate the underlying systems or processes. Subsequently also, we can produce simulated solutions for the system in regard to its predicted performance and examine the predicted output with actuals realized in future.

Monte Carlo simulation of chance-dependent systems, system dynamics to study how structures, policies and the timing of decisions and actions influence output of a complex feedback system and deterministic simulation to relate outcomes of a series of inputs which are produced by pre-determined decision rules – all illustrate the use of simulation.

With the advent of new technologies like underwater communication networks or exploration of groundwater or any special treatment of a material like tungsten, simulation, modelling and characterization of the new material or the new technology pose challenging problems for software development.

Monte Carlo simulation developed by S. Ulam and J. Von Neumann is used to model the probabilities of different outcomes in a process that cannot be easily predicted due to the intervention of random variables. This technique helps us to understand the impact of risk and uncertainty in prediction and forecasting models. Monte Carlo simulation can be used to tackle a wide range of problems from diverse fields such as finance, engineering, supply chain and science (Investopedia). Simulation here has to start with an assumed probability distribution of the outcomes and is based on generation of pseudo-random numbers, having recourse to the central limit theorem for estimation of the true value of the outcome.

Monte Carlo Markov chains provide useful tools for numerical integration dealing with complex integrands and have found applications in a wide array of Bayesian computations.

7.13 Model Uncertainty

Uncertainty may best be explained as a situation in which we have no unique, objective understanding of the system or process or phenomenon under study. And such a study usually starts with a model that represents or is expected to represent the pertinent features of the evidence available – generally by way of data – on the system or phenomenon. Needless to say that we are often uncertain about the appropriateness or adequacy of a model in representing the phenomenon of interest. Structural or model uncertainty arises primarily from lack of adequate – if not complete – knowledge about the underlying true physics of the phenomenon under study and the inadequacy of the measurements or observations on the phenomenon. There could be some bias on the part of the investigator that gets reflected in the model chosen. Also, discrepancies between the model chosen and the underlying mechanism that generates the measurements or observations may contribute to uncertainty For example, in the study of particle dynamics, the assumption of 'free fall' fails to recognize friction in air and may render any deductive inference based on a model that includes this assumption 'uncertain'.

Then comes parameter uncertainty due to inadequacy or inconsistency in the inputs to the model, particularly by way of values or levels of the parameters in the model which are not known uniquely, e.g. various material properties in a finite element analysis in engineering. Associated with the above is parameter variability over occasions – maybe within known ranges – that is sometimes caused by the nature of the phenomenon and its likely change over time or other domains. The above three taken together are broadly recognized as model uncertainty and this is what influences the general credibility of models as components in inference-making.

Of the many approaches to deal with model uncertainty, Bayesian analysis has been vigorously pursued in many scientific investigations and has been modified to incorporate model and model parameters both as random entities. Thus we first generate a model out of an existing kit of models, then associate some prior distribution with the parameter vector in the chosen model and then generate data from the chosen family.

A comprehensive Bayesian approach involving multiple model set-ups has sometimes been adopted to deal with the problem of model uncertainty by not only considering the parameters in a chosen model as random but also expanding the horizon of choice to incorporate several models altogether. In this expanded framework, we have the opportunity to explore the extent of uncertainty in a model, given the data, and to apply different routes to the basic problem of reducing uncertainty attributable to the 'model' in data analysis and data-based inferencing.

In this approach, we proceed to start with a set of models M_i, assigning a prior probability distribution $p(\theta_k \mid M_k)$ to the parameters of each model, and a prior probability $p(M_i)$ to each model. The set of models should, of course, share certain common properties that reflect the feature(s) of the underlying phenomenon of the enquiry or of the data available on the phenomenon. Thus, for example, dealing with failure-time data, the set of models to start with could be members of the NBU (New Better than Used) class if so suggested by the known behaviour of the equipment whose failure times have been observed. The priors for the parameters involved in the models initially considered may be the corresponding conjugate priors. In fact, it will be reasonable to use the same type of priors for the different models. This prior formulation induces a joint distribution $p(Y, \theta_k, M_k) = p(Y \mid \theta_k, M_k) \, p(\theta_k \mid M_k) \, p(M_k)$ over the data, parameters and models. In effect, we generate a large hierarchical mixture model.

Under this full model, the data are treated as being realized in three stages: first the model M_k is generated from $p(M_1)$, ..., $p(M_k)$; second the parameter vector θ_k is generated from $p(\theta_k \mid M_k)$; third the data Y are generated from $p(Y \mid \theta_k, M_k)$. Through conditioning and marginalization, the joint distribution $p(Y, \theta_k, M_k)$ can be used to obtain posterior summaries of interest.

Integrating out the parameters θ_k and conditioning on the data Y yields the posterior model probabilities

$$p(M_k \mid Y) = \frac{p(Y \mid M_k)p(M_k)}{\sum_k p(Y \mid M_k)p(M_k)}$$

where

$$p(Y \mid M_k) = \delta\int p(Y \mid ,_k, M_k)p(,_k \mid M_k)d,_k$$

is the marginal likelihood of M_k. (When $p(\theta_k \mid M_k)$ is a discrete distribution, integration in the equation above is replaced by summation.) Under the full three-stage hierarchical model interpretation for the data, $p(M_k \mid Y)$ is the conditional probability that M_k was the actual model generated at the first stage.

We can now introduce pair-wise comparison between models j and k on the basis of the posterior probabilities to eventually proceed towards choosing a model in terms of the posterior odds

$$\frac{p(M_k \mid Y)}{p(M_j \mid Y)} = \frac{p(Y \mid M_k)}{p(Y \mid M_j)} \times \frac{p(M_k)}{p(M_j)}$$

This expression reveals how the data, through the Bayes factor $B[k{:}j] \equiv \frac{p(Y \mid M_k)}{p(Y \mid M_j)}$, updates the prior odds $O[k{:}j] = \frac{p(M_k)}{p(M_j)}$ to yield the posterior odds. The Bayes factor $B[k{:}j]$ summarizes the relative support for M_k versus M_j provided by the data. Note that the Bayes posterior model probabilities can be expressed entirely in terms of Bayes factors and prior odds as

$$p(M_k \mid Y) = \frac{B[k:j]O[k:j]}{\sum_k B[k:j]O[k:j]}$$

In so far as the priors $p(\theta_k \mid M_k)$ and $p(M_k)$ provide an initial representation of model uncertainty, the model posterior $p(M_1 \mid Y)$, ..., $p(M_k \mid Y)$ provides a complete representation of post-data model uncertainty that can be used for a variety of inferences and decisions. By treating $p(M_k \mid Y)$ as a measure of the 'truth' of model M_k, given the data, a natural and simple strategy for model selection is to choose the most probable M_k, the modal model for which $p(M_k \mid Y)$ is the largest. Model selection may be useful for testing a theory represented by one of a set of carefully studied models, or it may simply serve to reduce attention from many speculative models to a single useful model. However, in problems where no single model stands out, it may be preferable to report a set of models, each with a with high posterior probability, along with their probabilities to convey the model uncertainty.

Bayesian model averaging is an alternative to Bayesian model selection that accepts rather than avoids model uncertainty. For example, suppose interest is focused on the distribution of Y_r, a future observation from the same process that generated Y. Under the full model for the data induced by the priors, the Bayesian predictive distribution of Y_r is obtained as

$$p(Y_r|Y) = \sum_k p(Y_r|M_k, Y)p(M_k|Y)$$

which is virtually a posterior weighted mixture of the conditional predictive distributions

$$p(Y_r|M_k, Y) = \int p(Y_r|_{,k}, M_k)p(_{,k}|M_k, Y)dk$$

By averaging over the unknown models, $p(Y_t | Y)$ incorporates the model uncertainty embedded in the priors in a way that only circumscribes model uncertainty within a certain domain. A natural point prediction of Y_r is obtained as the mean of $p(Y_r | Y)$, namely

$$E(Y_r|Y) = \sum_k E(Y_r|M_k, Y)p(M_k|Y)$$

Such model averaging or mixing procedures to contain model uncertainty have been developed and advocated by Leamer (1978), Geisser (1993), Draper (1995), Raftery, Madigan and Volinsky (1996) and Clyde, DeSimone and Parmigiani (1996).

A major appeal of the Bayesian approach to model uncertainty is its complete generality. In principle, it can be applied whenever data are treated as a realization of random variables, a cornerstone of model statistical practice. The past decade has seen the development of innovative implementations of Bayesian treatments of model uncertainty for a wide variety of potential model specifications. Of course, different ways of selecting one among the models with which we initially intend to represent the data based on different selection criteria will be there. Sometimes, the choice should depend on the objective of selecting a model as an intermediate step in the quest for a final answer to a not-so-simple question. Each implementation has required careful attention to prior specification and posterior calculation.

One may even look at the data part in this context. The data would yield realized values of some random variables and the choice of these variables as reflected in the data does have a significant role in the model specification and selection exercise. In fact, one wonders if the Bayesian approach does provide any assistance to the research worker in the choice of variables which can minimize the extent of model uncertainty at the next stage and can lead to robust conclusions later.

The key object provided by the Bayesian approach is the posterior quantification of post-data uncertainty. Whether to proceed by model selection or model averaging is determined by additional considerations that can be formally motivated by decision theoretic considerations (Gelfand, Dey and Chang 1992; Bernardo and Smith, 1996) by introducing the concept of a quantifiable utility or a quantifiable loss. Letting u(a, Δ) be the utility or negative loss of action a given the unknown entity of interest Δ, the optimal a maximizes the posterior expected utility

$$\int u(a, \Delta) p(\Delta \mid Y) d\Delta,$$

where $p(\Delta \mid Y)$ is the predictive distribution of Δ given Y under the full three-stage model specification. Thus, the best posterior model selection corresponds to maximizing $0 - 1$ utility for a correct selection. The model averaged point prediction $E(Y_r \mid Y)$ corresponds to minimizing the quadratic loss with respect to the actual future value Y_r. The predictive distribution $p(Y_r Y)$ minimizes Kullback-Leibler loss with respect to the actual predictive distribution $p(Y_r \mid \theta_k, M_k)$.

8

Data Analysis

8.1 Introduction

As can be easily appreciated, data analysis using cognitive ability to argue logically with facts and figures, to visualize and summarize the data after checking their relevance and validity and also credibility, to extract desired information and derive pertinent knowledge from the data by subjecting them to appropriate qualitative and quantitative tools using algorithms and software whenever necessary and finally interpreting results of such analyses in terms of the objectives of research is the key task that occupies a majority of the time and effort of the research worker.

It must be remembered that methods and tools for analysis of research data have to take due account of the type of data to be collected (as indicated in the research design) and any special features of the data (as could be eventually collected). And we also notice that a huge volume of empirical research makes use of secondary data gathered from different sources where we have little control over the mechanism used in these sources. Special features not often foreseen during the design phase including missing entries or gaps in a time series, data censored in various ways, data suspected to be outliers, etc.

Research data analysis should not be dictated by familiarity or knowledge of the research worker in some methods and tools of analysis – simple or sophisticated, traditional or recently developed. It is, of course, true that for the same intended task a complex method or tool should not be avoided in favour of a simple one and a recently developed method or tool should be preferred to a classic one. It is also true that some standard methods and tools for data analysis should be known to all research workers.

Any method of analysis will involve some model or other and such a model can always incorporate any previous information or knowledge about different elements in the model. The simplest example could be the use of any prior knowledge about the parameter(s) in the distribution of some random variable assumed in the model, by assuming such a parameter as a random variable having a prior distribution with given values of the parameters as in the Bayesian set-up. The idea can be extended further to the hierarchical Bayes model.

The above points will be clear if we examine the objective(s) and the nature of data likely to arise in the context of the following research questions, taken severally.

a. To identify some optimum stocks/assets in which to invest money and to work out amounts of stocks to be purchased on a consideration of prevailing market returns and risks. One may work with multiple time series of vector auto-regressive type, introduce vector error correction, take into account co-integration and solve a quadratic programming model for the purpose.

b. To study the problem of passenger congestion in a metro rail network and find the optimal frequency of service, one may use a state-space model reflecting the patterns of traffic flow in clusters of stations like those located in business areas, residential blocks and educational institutions etc. and make use of a recently developed method called statistical transfer learning.

c. To examine variations in social exclusion (inclusion) in different regions, we may consider several indicators of exclusion and determine appropriate weights to work out a single index which will allow fair comparison across regions. One may possibly think of data envelopment analysis for developing these weights endogenously based on the data collected on the indicators for the regions. One can introduce the concept of benefit-of-doubt to avoid the backward regions in poor light.

d. To analyse trends in pollution over a certain region in terms of several parameters, it is quite possible to recognize long gaps in data on some of the parameters, and one could think of functional data analysis and empirical orthogonal functions to impute the missing entries.

Multivariate (data) analysis corresponds to situations – quite commonly come across in empirical research – where multiple characteristics (measurable/countable/observable) are noted in respect of each experimental/observational unit and are considered together to provide the data for making inferences about the phenomena/processes being investigated.

With advances in science and, particularly, in information and communication technology, more and more characteristics are being noted to get deeper and closer insights into the phenomena or processes under study. Quite often, such characteristics are recorded automatically and also instantaneously.

In this chapter, we concentrate mostly on data generated by the researcher either through a sample survey or as responses of experimental units in some designed experiments. And, for the sake of simplicity, we restrict ourselves to univariate data. A somewhat detailed discussion on analysis of multivariate data is postponed to the next chapter. It has to be kept in mind that missing values of some variable(s) in the study are missing. Ignoring such values may not provide a correct understanding of the underlying phenomenon and we may have to impute them by following a suitable algorithm. The pattern revealed by the missing values, viz. missing completely at random, missing at random and missing not randomly, has to be taken account of in determining the proper treatment of missing values. In the multivariate case, if quite a few missing values occur those should just be deleted, while in univariate analysis we may think of imputation of missing values.

Some of the characteristics are directly measured, while certain others are derived from components or proxies/substitutes. The word 'unit' need not necessarily correspond to single living beings or non-living material objects; units could signify groups of constant or varying sizes of such beings or objects.

A properly designed experiment or a survey with a proper sampling design and survey instrument or a mechanism to compile secondary data is meant to provide inputs into a research process and the corresponding research results through analysis of the data. And this data analysis may involve several methods or models and tools all together. No doubt, individual methods, models and tools have their roles and need to be understood by the research worker. However, it will be proper if we avoid discussing these methods and tools on a stand-alone basis. Instead, analysis of a few data sets arising in the context of some research has been portrayed in the following.

To conclude, we should note that data analysis in the context of some research or some part thereof does not mean choosing and applying a particular method of analysis or the associated tools to the data collected as are relevant. Whatever methods and tools are required to extract the information we want to meet the objectives of the research may have to be invoked, taking due care about the validity of their applications in the given context. And this is what has been attempted in the following sections.

8.2 Content Analysis of Mission Statements

Many educational institutions develop and display their mission statements at different prominent locations within their premises. In particular, we may focus on schools and try to find out if privately funded schools give more emphasis to emotional or physical development of their students compared to publicly funded institutions, where supposedly more emphasis is laid on academic development. Had a survey been planned to set opinions from a group of teachers or of students or of parents/guardians, responses would be quite possibly biased by the backgrounds of the respondents. On the other hand, we can reasonably consider the mission statements as reflecting the emphasis placed by a school on academic, emotional and physical development, though actual practice could differ from the content of the mission statement.

Assuming the mission statement to be representing the ground reality, we take a sample of, say 100, privately funded schools and request two experts to go through the 100 mission statements to find out which of the three developments – physical, emotional and academic – gets the highest emphasis in a mission statement. The experts may disagree between them, even when they have been brought to the same table to follow some agreed procedure to identify the aspect most emphasized. While it is admitted that two or all three aspects are emphasized in a mission statement, we would like the expert to identify the most emphasized aspect. The results came out as shown in Table 8.1.

Experts have been mentioned as raters, as is the practice in content analysis.

Reliability may be calculated in terms of agreement between raters using Cohen's kappa, lying between 1 to imply perfect reliability and 0 indicating agreement only through chance, and given by

$$K = (P_A - P_C) / (1 - P_C)$$

where P_A = proportion of units on which raters agree and P_C = proportion of agreement by chance.

Here $P_A = 0.42 + 0.25 + 0.05 = 0.72$ and $P_C = 0.29 + 0.13 + 0.01 = 0.43$. So that Cohen's kappa = $(0.72 - 0.43) / (1.00 - 0.43) = 0.51$.

It may be noted that per cent agreement in this example is 72.

To interpret this value of K obtained from the data, we refer to the following table due to Landis and Koch (1977) – based on personal opinion and not on evidence. These guidelines may be more harmful than helpful, as the number of categories and of subjects will affect the value of Kappa. The value will be higher with fewer categories.

TABLE 8.1

Aspects Most Emphasized in Mission Statements

Rater 1		Marginal			
		Academic	Emotional	Physical	Marginals
Rater 2	Academic	.42 (.29)	.10(.21)	.05 (.07)	.57
	Emotional	.07 (.18)	.25(.13)	.03 (.05)	.35
	Physical	.01(.04)	.02(.03)	.05(.01)	.08
	Marginal	.50	.37	.13	1.00

TABLE 8.2

Aspect of Development vs. Type of Funding

Aspect of Development	Type of Funding		
	Private	Public	Total
Academic	54	99	153
Emotional	36	21	57
Physical	10	20	30
Total	100	140	240

Value of K	Strength of agreement
< 0	poor
.00–.20	slight
.21–.40	fair
.41–.60	moderate
.61–.80	substantial
.81–1.00	almost perfect

Therefore, we can possibly infer a moderate agreement between the two raters in this example. This enables us to consider the average of the two marginals as representative of the situation in privately funded schools and this is (0.54, 0.36, 0.10). We could have checked the agreement among three or more raters as well.

Similarly, we got the figures for 140 publicly funded schools and the kappa coefficient of agreement between the same two raters came out as 0.74, which is 'substantial'. However, the vector agreed upon came out as (0.71, 0.15, 0.14).

The researcher may be interested to test the hypothesis that "the most emphasized aspect of development is independent of the type of funding". This hypothesis implies no association between the two classifications, viz. type of funding and the most emphasized aspect of development. Table 8.2 shows the two-way table of data for this purpose.

To test for independence, we use the Brandt-Snedecor formula to calculate the chi-square statistic, viz. $\chi^2 = [\Sigma\, a_i^2 \,/\, T_i - T_A^2 \,/\, T]\, T^2 \,/\, T_A\, T_B = 14.22$ with 2 degrees of freedom. The critical value of χ^2 at 1% level of significance is 9.21 and thus the observed value is highly significant, implying that the hypothesis of independence is rejected. In operational terms, it implies significant difference in respect of the most emphasized aspect of students' development between privately funded and publicly funded schools. In fact, the data reveal that unlike publicly funded schools which give the highest emphasis to academic development, private schools put more stress on physical and emotional development.

8.3 Analysis of a Comparative Experiment

An experimenter is examining the effects of five different formulations of an explosive mixture to be used in the manufacture of dynamite. Each formulation is mixed from a batch of raw material that is just large enough for five formulations to be tested. Further, the formulations are prepared by several operators who may have substantial differences in skill and experience. The design appropriate for this experiment should be:

> test each formulation exactly once from each batch of raw material and each formulation to be prepared exactly once by each operator. This will eliminate the extraneous influences of two factors, viz. the batch of raw material and operator skill and experience, on the comparison among the five formulations in respect of the response variable, say, depth of crater created on use of the dynamite. This becomes a Latin square design with rows as raw material batches and columns as operators.

Table 8.3 gives the layout of the experiment and also the responses
The five formulations have been denoted by the letters A, B, C, D and E. The linear model for analysing variations in this experimental set-up is given by $Y_{ijk} = \mu + \alpha_i + \beta_j + \lambda_k + \varepsilon_{ijk}$ (i, j. k) ε D where D stands for the design in terms of the triplets actually used in the set-up, μ is the general effect, α_i is the fixed specific effect of row (raw material batch) i, β_j is the fixed specific effect of column (operator) j and λ_k is the effect of formulation k, while ε_{ijk} is the unexplained random error. The unknown parameters of interest α_i, β_j and λ_k satisfy the linear relation $\Sigma\, \alpha_i = \Sigma\, \beta_j = \Sigma\, \lambda_k = 0$ and the random errors are independently normally distributed with mean 0 and variance σ_e^2.

The analysis of variance (ANOVA) is presented in Table 8.4.

TABLE 8.3

Showing the Layout and the Responses in an LS Experiment

		Operators				
		1	2	3	4	5
Batch of Raw	1	A (24)	B (20)	C (19)	D (24)	E (24)
Material	2	B (17)	C (24)	D (30)	E (27)	A (36)
	3	C (18)	D (38)	E (26)	A (27)	B (21)
	4	D (26)	E (31)	A (26)	B (23)	C (23)
	5	E (22)	A (30)	B (20)	C (29)	D (31)

TABLE 8.4

ANOVA for the Latin Square Experiment

Source of Variation	Degrees of Freedom	Sum of Squares	Mean Squares	F-ratio
Batches	4	68.00	17.00	
Operators	4	150.00	37.50	
Formulations	4	333.00	82.50	7.73
Error	12	128.00	10.67	
Total	24	676.00		

The critical value of the F-ratio with 4 and 12 degrees of freedom at the 5% level is 3.26. Thus the differences among the five formulations come out to be significant at the 5% level. One may now feel interested in pair-wise comparisons. Incidentally, the mean responses of the five treatments A, B, C, D and E are respectively 28.6, 20.2, 22.4, 29.8 and 26.0. It comes out that formulations A and D provide large responses, followed by formulation E, while the other two formulations have lower responses. We may like to test whether there is any significant difference between A and D by using the t-test. The computed value of the t-statistic will be t = (29.8 − 28.6) / $\sqrt{}$ [2 MSE / 4] = 1.2 / $\sqrt{}$ 5.33 = 0.225 and this is less than the 5% critical value of t with 4 d.f., implying no significant difference between formulations A and D. However, on comparing formulations A and C we get the t-statistic computed as t = (28.6 − 22.4) / $\sqrt{}$ [2 MSE / 4] = 6.2 / 2.3 = 2.70 and this is significant at the 5% level.

8.4 Reliability Improvement through Designed Experiment

Properly designed experiments with levels of factors which influence the reliability of the system/product, coupled with an exploration of the response surface, yield the optimum design for the best attainable reliability. Responses in such experiments are time-to-failure (maybe with a censoring plan) along with some other functional requirement(s) which may enter by way of some constraint(s).

The life test in this connection could be terminated after a certain specified time, say t_0, usually smaller than the mission time up to which the product is expected to function satisfactorily. In the simplest case, the product either survives till t_0 or fails earlier, when the only performance parameter, viz. time-to-failure, drops below the specified value. In this case, we have to monitor the performance parameter(s) continuously and record time-to-failure as defined above. Alternatively, we simply record the time of functioning of the product till the censoring time t_0.

After examining the dependence of survival time on the design parameters, we are required to optimize the choice of parameters by maximizing reliability subject to constraints on the performance parameter(s). In fact, we should (a) find a suitable distribution of time-to-failure and (b) fit a regression equation of the scale parameter on the design parameter. For this, we have to (c) estimate the scale parameter of the fitted distribution for each design point based on its few replications.

- While the simple one-parameter exponential with the p.d.f. f (x) = λ exp (− λx) is an obvious choice, it may be better to try out the more general Weibull distribution with p.d.f. f (x; α, λ) = $\alpha\lambda$ x α − 1 exp (− λ xα).
- In the latter case, it may not be unreasonable to assume that the shape parameter α remains the same across design points and we consider the dependence of the scale parameter λ only on the design points. We estimate α based on all the design points and values of λ for each point.

Consider the example given in Reliasoft (2014). Reliability required for the device under experimentation (in a water supply system) = 0.95 for a volume of 80 gallons delivered per hour (acceptable range 75–85 gallons) and survival time for at least 2000 hours.

The controllable factors (during design and manufacture) in the product design affecting volume of water are

Thickness (from 2 to 6 inches)
Tuning time (from 10 to 50 minutes)
Width (from 8 to 18 inches)

A central composite design with two levels (high and low) for each of these three factors was chosen to facilitate the exploration of a quadratic response surface.

Thickness u (5 and 3 inches)
Tuning time v (40 and 20 minutes) and
Width w (15 and 19 inches) used to note (1) time-to-failure (t) and (2) gallons / hour (y) in a test censored at 2000 cycles
Response function for y (gallons/hour) taken as

$$y = \alpha + \beta_1\, u + \beta_2\, v + \beta_3\, w + \beta_{12}\, uv + \beta_{13}\, uw + \beta_{23}\, vw + \beta_{11}\, u^2 + \beta_{22}\, v^2 + \beta_{33}\, w^2 + \varepsilon$$

Apart from the 23 = 8 usual factor-level combinations (vertices of the cube spanned by the eight treatment combinations), six axial points with low and high α values taken as thickness (2.3, 5.7) tuning time (13.2, 46.8) and width (8.3,16.7) were each replicated twice, while six replications were done for the central point (4, 30, 12.5).

In all 34 runs of the experiment were carried out with 15 design points and the two responses noted for each run.

ANOVA table for 34 runs and the model chosen will have 1 d.f. for each main effect U, V and W, each first-order interaction UV, U W and V W and each quadratic effect U^2, V^2 and W^2, plus 24 d.f. for residual (including 5 for lack of fit).

The p-values in the ANOVA table show that V, W, U W and W2 are significant.

And the revised regression (with U and these four) comes out as

$$y = 469.8 - 55.1U - 0.2\, V - 56.7\, W + 4.4\, U\, W + 1.88W2 + \varepsilon$$

Examining the response surfaces, the best settings for the three factors come out as

Thickness = 2 inches
Tuning time $\approx$ 39 min
Width $\approx$ 17 inches

For the purpose of optimizing reliability, we need to consider a probability model for time-to-failure. For modelling time-to-failure (t), we assume t to follow (a) an exponential distribution, (b) a Weibull distribution and (c) a log-normal distribution. We try these models sequentially.

8.4.1 Exponential Failure Model

Suppose time-to-failure t is assumed to have a one-parameter exponential distribution with density function $f(x; \lambda) = \lambda\, e^{-\lambda x}$. The mean failure time is given by $1\,/\,\lambda$. We therefore model λ as a function of the three factors (u, v, w), and the relationship is given by $\lambda = \exp(- \mu)$, where

$$\mu = b_0 + b_1\,u + b_2\,v + b_3\,w + b_{11}\,u^2 + b_{22}\,v^2 + b_{33}\,w^2 + b_{12}\,u\,v + b_{13}\,u\,w + b_{23}\,v\,w$$

Based on the observed data, the log likelihood function is obtained as

$$\text{Log } L = -\, \Sigma_1\mu_i - \Sigma_1\, t_i \exp(- \mu_i\, t_i) - (2000)\, \Sigma_2 \exp(- \mu_i\, t)$$

where
Σ_1 = summation over all i for which $t_i < 2000$
Σ_2 = summation over all i for which $t_i > 2000$.

For all calculations, we use the interior point algorithm in the software MATLAB.
The MLE of the regression of ln λ on the design parameters is obtained as

$$\ln \hat{\lambda} = -[0:8594 - 0:6178u + 3:0795v - 5:5910w - 3:5256u^2 - 0:0251v^2 +$$
$$0:44021w^2 - 0:1155uv + 2:0817uw - 0:1780vw]$$

Corresponding to the optimum setting of the factors, as obtained from the revised regression of volume (y) on u, v and w, the predicted value of λ is $1{:}46041 \times 10^{-19}$, and the reliability at 2000 hours is approximately 1.

The problem of improving reliability would be to maximize the reliability at 2000 hours, namely exp (– 2000λ), with proper choice of the design parameters u, v and w, lying in their respective ranges given by $2 \le u \le 6$, $10 \le v \le 50$, $8 \le w \le 18$, subject to the volume (y) being at least 80 gallons per hour.

The optimum choices of the design parameters come out to be: thickness= 2.72 inches, tuning time =35.91 mins and width = 16.69 inches, and the reliability at $T = 2000$ is approximately 1, with the volume 80 gallons /hour.

8.4.2 Weibull Failure Model

Here we assume that t follows a Weibull distribution with p.d.f. given by

$$\lambda = \exp\left(-\frac{\mu}{\sigma}\right),$$

where

$$\sigma = 1/\,\alpha,$$

$$\mu = a_0 + a_1u + a_2v + a_3w + a_{11}u^2 + a_{22}v^2 + a_{33}w^2 + a_{12}uv + a_{13}uw + a_{23}vw.$$

The log-failure-time is then

$$\log_e t = \mu + \sigma\varepsilon,$$

where ϵ is the error term distributed with p.d.f.

$$g(\varepsilon) = \exp[\varepsilon - \exp(\varepsilon)], \quad -\infty < \varepsilon < \infty.$$

Based on the data in Table 4.1, and the assumption that α remains the same across all design points, the log likelihood function comes out to be

$$\text{Log } L = 34 \ln \alpha + |(\alpha - 1)\Sigma_1 \ln(t_i) - \alpha\Sigma_1\mu_i - \Sigma_1 t_i^{\alpha} \exp(-\alpha\mu_i) - (2000)^{\alpha}\Sigma_2 \exp(-\alpha\mu_i)$$

where

$$\Sigma_1 = \text{summation over all i for which } t_i < 2000$$
$$\Sigma_2 = \text{summation over all i for which } t_i > 2000.$$

The MLE of α comes out to be $\hat{\alpha} = 4.525$, and the estimated regression of $\log_e\lambda$ on the covariates is obtained as

$$\ln \hat{\lambda} = -[0.0634 - 2.6743u + 0.9593v + 3.5069w - 0.1403u^2 - 0.0136v^2 + 0.1448w^2 \\ - 0.0136uv + 0.1131uw + 0.0091uw] \tag{5.2}$$

The optimum setting of the factors obtained from the revised regression of volume (y) on u, v and w gives the predicted value of λ as 5:30101 × 10^{-17}, and the reliability at 2000 hours is 0.9552.

Here the problem of improving reliability would be to maximize the reliability at 2000 hours, namely exp (− 2000λ), with proper choice of the design parameters u, v and w, lying in their respective ranges given by $2 \le u \le 6$, $10 \le v \le 50$, $8 \le w \le 18$, subject to the volume (y) being at least 80 gallons per hour.

The optimal choice of the design parameters are obtained as thickness = 2 inches, tuning time =34.063 mins and width = 8.593 inches, and the reliability at 2000 hours is 1 with volume 80 gallons/hour.

8.4.3 Lognormal Failure Model

Suppose the time-to-failure t is assumed to have a lognormal (μ, σ^2) distribution.

The expected time-to-failure is then given by E (T) = exp ($\mu + \sigma^2 / 2$), which is an increasing function of the location parameter μ. Hence, assuming μ to depend on the design points while the shape parameter σ remains constant, we model μ as follows:

$$\mu = b_0 + b_1u + b_2v + b_3w + b_{11}u^2 + b_{22}v^2 + b_{33}w^2 + \\ b_{12}uv + b_{13}uw + b_{23}vw$$

The log-likelihood function then comes out to be

$$\text{Log } L = -(m/2)\ln(2\pi) - (m/2)\ln(\sigma^2) - \Sigma_1 \ln(t_i) - (1/2\sigma^2)\Sigma_1(t_i - \mu_i)^2 \\ + \Sigma_2 \ln \Phi\big((\ln(2000) - \mu_i)/\sigma\big)$$

where
Σ_1 = summation over all i for which $t_i < 2000$
Σ_2 = summation over all i for which $t_i > 2000$
m = number of observations with $t_i < 2000$
Φ (.)= cumulative distribution function of a standard normal variate with

$$\Phi(-t) = 1 - \Phi(t)$$

The MLE of the fitted regression of μ on the covariates comes out to be

$$\hat{\mu} = 0:0958 + 0:1818u + 0:2317v + 0:3376w + 0:4179u^2 + 0:0095v^2$$
$$+ 0:2337w^2 + 0:4839uv + 0:6302uw + 0:1665vw;$$

and the MLE of σ is 3.4138. The optimum values of the covariates that maximize the reliability at T = 2000, subject to the constraint that the volume is at least 80 gallons/hour, are estimated to be thickness = 10.92 inches, tuning time =29.9 mins and width = 10.49 inches. The corresponding estimate of the reliability at T = 2000 is approximately 1, with the volume 80 gallons/hour.

It thus comes out that we can achieve a reliability of almost 1 with proper choices of the design parameters, whichever model to represent random variations in failure time is selected. However, the design points will not remain identical under the three set-ups.

8.5 Pooling Expert Opinions

In several research investigations, one has to make a choice among alternative decisions or courses of action or even some objects or subjects judged by some criterion or criteria used by a group of experts. These experts have to be chosen in a manner that suits the context, may be allowed to interact among themselves fully or partly or may not be allowed to interact, may be required estimates in one go or allowed to revise on the basis of some feedback provided from the previous round of exercise till there is sufficient convergence. Estimates spoken above may mean ranks assigned to the different entities considered or paired comparisons leading to one of the two entities being preferred to the other or some distance measure between the two members or even probabilities of an event likely to occur in future.

There are different situations in which experts act and different procedures for choice between any two or among all the candidates may be followed. Accordingly there are different ways of making using of expert opinions (including an assessment of the extent to which experts agree among themselves) and arriving at some numerical basis for making a choice among the candidates. In this context, we will consider here only one or two situation-procedure combinations, just to illustrate the overall problem.

8.5.1 Delphi Method

The Delphi method is the best-known procedure with feedback. Experts are asked their answers to several questions, usually along with the arguments in their favour. The analyst studies the answers of experts and finds their points of agreement. If the expert opinions are not in sufficient agreement, the analyst gives to each expert more data about the system and also answers and supporting arguments given by the other experts. The experts give their revised opinions on the basis of the additional data provided. An important drawback is the excessive time that may be required to ensure convergence or sufficient agreement among the expert opinions to allow averaging.

Take the example (Makarov et al., 1987) where ten experts are required to estimate an unknown parameter and the following estimates are obtained: 35, 35, 32.2, 34, 38, 34, 37, 40, 36 and 35.5 yielding an average of 35.5 and a standard deviation $\sigma = 2.2136$. Thus, we can claim that the interval (33.917, 37.083) contains the true parameter value with a probability 0.95. Now think of a Delphi exercise where the experts are provided with the median value of the estimates and the inter-quartile range along with justifications for some estimates falling outside this range. The interval of feasible values for the estimated parameter in divided into k sub-intervals and each expert is to estimate the probability that the parameter will lie in each of the sub-intervals. Assume that p_{ij} is an estimate of the probability that the parameter will lie in the jth sub-interval in the opinion of the ith expert. We can then take the probability that the parameter will lie in the jth sub-interval as $\Sigma \, \alpha_i \, p_{ij} \, / \, \Sigma \, \alpha_i$ where α_i represents the weight assigned to expert i. Most often these weights are equal.

It is quite imaginable that the experts were isolated one from the other and, at the end of one round in which their estimates have been noted, feedback in terms of the median value and the inter-quartile range can be provided. This procedure can be repeated until the range becomes pretty small and the number of estimates/opinions outside this range is very small.

8.5.2 Analysis of Rankings

In many research studies, we deal with ranks assigned to a set of n units or individuals by some experts or judges, which may be based on several criteria or features possessed by the units. In some applications of ranking, we may decide to choose the best or the few better than all the rest. The reverse could also be the case, for example when the ranks are used to identify the poorest or the worst units or units which are poorer or worse than the remaining, so that these units may be removed from further consideration or are referred to some 'recovery treatment' ranking that can be done in different procedures and the features observed for each unit may be summarized or converted to some indicator in several possible ways. It is also possible that ranks are first obtained in respect of each criterion or feature and subsequently the average ranks are considered. All the units could be presented before an expert or judge to be ranked from 1 (the best) to n (the worst), avoiding ties. Alternatively, pairs of units may be presented to an expert who prefers one to the other within a pair. Assuming consistency of choice, we can eventually get ranks of all the n units. Ranks assigned by different judges may be tested for their agreement by calculating the coefficient of concordance, and if this measure if high we can justifiably average their ranks for each unit.

One problem of interest to many investigators concerns the comparison of efficiency among several 'decision-making units' or 'DMUs' which utilize some inputs to derive some outputs. These units could be different branches of the same bank or different campuses of the same school or could just be comparable units which deploy the same inputs to come up with the same outputs in different quantities. To compute efficiency we make use of data envelopment analysis (DEA), which involves repeated application of linear programming to derive weights accorded to different inputs and outputs and thus the efficiency of a DMU defined as the ratio between the weighted aggregate of outputs and the weighted aggregate of inputs. And we can rank the DMUs in terms of the efficiency. Of course, DEA has several variants and we have to choose one. Dimitrov (2014) makes an interesting study of DEA analysis and human rankings of DMUs, using an online survey to elicit the DMU rankings of 399 individuals from India. He considers four data sets, viz. (1) a hospital data set with 12 DMUs, two input and two outputs, (2) data relating to United Kingdom accounting departments with 12 DMU's, three input and seven output variables, (3) airline data with six DMUs involving six input and five output variables and (4) power-generation data from six DMU's with three input and one output variable.

Dimitrov makes use of several DEA methods, viz.:

The super-efficient DEA approach by Anderson and Peterson

The bounded adjusted measure of efficiency defined by Cooper et al.

The canonical correlation analysis–DEA approach defined by Friedman and Sinuany-Stern

The aggressive cross-efficiency model defined by Doyle and Green

The benevolent cross-efficiency DEA model proposed by Doyle and Green

The Charnes, Cooper and Rhodes DEA model

The additive DEA model proposed by Charnes et al.

The benchmark DEA model proposed by Torgerson, Forsund and Kittelsen

The principal-component analysis DEA model as proposed by Adler and Golany

The SWAT or symmetric weight assignment technique proposed by Dimitrov and Sutton

For the first data set with only two inputs and two outputs, it is expected that each participant in the survey selected from persons with collegiate education would be able to rank the 12 DMUs, maybe using tied ranks in some cases. The authors quote the ranks of the first ten participants and these are found to have some tied ranks. It is worthwhile to examine the extent of agreement among these ten respondents and for this we may calculate Kendall's coefficient of concordance W, essentially based on ANOVA of the ranks between and within participants. Using the formula $W = 12 S / [n (n^2 - 1) k^2]$ where S stands for the sum of squared deviations of rank totals of the DMUs from their mean, n is the number of DMUs, viz. 12 in this case, and k is the number of participants. We have to replace each tied rank by the mid-rank and adjust ranks of other DMUs accordingly.

Table 8.5 gives the ranks assigned to the 12 DMUs by the first ten participants as reported by the authors.

It is amply clear that the participants differ a lot in the ranks they assign to a particular DMU, primarily because of differences in the way they combine the output and input values to arrive at a measure of efficiency. In fact, the same DMU has received the lowest and also the highest rank in quite a few cases. Further, some of the participants cannot distinguish between two or more DMUs so that we have many tied ranks. In fact, no ties are observed in the ranks assigned by participants 1, 2, 3, 4 and 10. The formula for computing the coefficient W in the case of tied ranks is given by

$$W = [12 S' - 3 k^2 n (n + 1)^2] / [k^2 (n^3 - n) - k T]$$

TABLE 8.5

Ranks Assigned to 12 DMUs by 10 Participants

	DMU											
Participant	A	B	C	D	E	F	G	H	I	J	K	L
1	11	8	6	9	3	2	1	7	12	4	5	10
2	10	12	11	8	9	1	5	7	6	3	2	4
3	6	12	8	1	7	2	9	11	5	3	4	10
4	10	12	11	8	9	2	5	6	7	1	3	4
5	4	5	5	2	1	3	6	7	8	8	11	10
6	11	11	9	8	10	6	4	3	5	7	1	2
7	1	5	5	1	2	5	4	4	2	2	3	4
8	10	10	9	7	9	6	5	8	4	3	1	2
9	2	1	5	4	2	2	3	3	3	6	5	3
10	8	9	10	11	12	5	4	7	6	2	1	3
Totals	73	85	74	59	64	34	46	63	58	39	36	52

where T is a correction factor for tied ranks given by $T = \Sigma \, (m_j^3 - m_j)$ summed over the g groups of ties, in which m_j is the number of tied ranks in each (j) of the g groups. S' is the sum of squares of total ranks received by the DMUs. In the above case, S' = 41.733, T = 1. Hence, the value of W comes out as W = 0.1853 and the corresponding value of chi-square with 11 degrees of freedom is 19.883, and this is highly significant, implying significant differences among the 12 DMUs in regard to their ranks as given by the ten participants. This really indicates that ranks assigned by the different participants for any DMU should not be averaged.

8.6 Selecting a Regression Model

Selecting an appropriate model within the relevant class based on some given observations is an important step in data analysis for research purposes. We have several criteria for choosing a model, essentially based on penalized likelihood. In the case of linear regression analysis, some function of the residual sum of squares (ascribable to deviation from linear regression) becomes the criterion. In fact, in this case, the different model selection criteria generally used are the following:

$$\text{Akaike information criterion AIC} = n. \ln (RSS) + 2k - n \ln(n)$$
$$\text{Bayes information criterion BIC} = n. \ln (RSS) + \ln (n) - k$$
$$\text{Mallow's } C_p = RSS / s^2 - n + 2p$$

where n = number of observations, k= number of explanatory variables p = number of explanatory variables in a sub-model, RSS = residual sum of squares, s^2 = estimate of variance of the dependent variable derived from the full model. In addition to these three criteria, we can also consider the adjusted R^2 (coefficient of determination).

We consider the data analysed by Rao and Shanbhag (1998) with 30 observations on y = expenditure on cultural activities per year (in Euro), x_1 = age (in years) and x_2 = gender (female =1, male = 0).

Sl.	y	x_1	x_2	Sl.	y	x_1	x_2
1	170	62	0	16	130	43	1
2	140	60	1	17	150	53	0
3	140	60	0	18	130	30	1
4	160	55	1	19	110	24	0
5	140	42	0	20	180	59	0
6	120	26	0	21	120	21	0
7	140	56	0	22	110	25	0
8	120	28	1	23	110	55	1
9	180	25	1	24	160	57	0
10	110	30	1	25	120	22	1
11	150	30	1	26	130	46	1
12	150	62	0	27	180	50	0
13	170	49	1	28	110	28	1
14	125	26	0	29	120	38	0
15	189	64	0	30	110	26	0

Five different linear regression models were fitted to the data, viz.:

$$\text{Model} \quad 1 \quad y = \beta_0 + \varepsilon$$
$$2 \quad y = \beta_0 + \beta_1 x_1 + \varepsilon$$
$$3 \quad y = \beta_0 + \beta_1 x_2 + \varepsilon$$
$$4 \quad y = \beta_0 + \beta_1 x_1 + \beta_2 x_2 + \varepsilon$$
$$5 \quad y = \beta_0 + \beta_1 x_1 + \beta_2 x_2 + \beta_{12} x_1 x_2 + \varepsilon$$

The authors reported the following values for the five models in respect of the three model selection criteria.

Model	AIC	BIC	C_p	R^2 (adjusted)
1	279.6527	282.4551	23.3164	
2	266.2639	270.4675	4.7242	0.3799
3	281.1725	285.3761	24.5016	0.0000
4	268.2636	273.8683	6.7238	0.3569
5	265.2553	272.2613	4.0000	0.4349

It appears that model 5 with only 'age', 'gender' and the interaction term as the explanatory variables is preferred by AIC, C_p and R^2 criteria, while BIC chooses the more parsimonious model 2 with only 'age' as the regressor. This, of course, is not very surprising, since model 5 exhausts more available information, compared to the other models. In a regression analysis, one often attaches greater importance to adjusted R^2 for choosing a model.

8.7 Analysis of Incomplete Data

Values of some variables may be missing in respect of some units for a variety of reasons in a sample survey or in a controlled experiment or even in some exercises to study the dependence of some response variable on some factors in some econometric or biological study. A detailed account appears in Rao et al. (2008). Rubin (1976, 1987), Little and Rubin (2002) and several others have contributed significantly to the analysis of incomplete data. Such occasions arise if

(a) respondents or informants in a survey do not provide information on certain items due to ignorance or wilful avoidance of some sensitive items or lack of interest in the survey;

(b) individuals withdraw themselves from long-duration clinical trials after some time, beyond which observations on them are not available;

(c) yield or some other response corresponding to a given treatment (treatment combination) may be destroyed or missing otherwise;

(d) in physical experiments involving some destructive test, values of some parameters will not be observable once the unit is destroyed or even damaged or mutilated.

As noted earlier, we may face situations where we have no response from a particular unit (individual or household or enterprise) and we have a completely missing unit (complete non-response in a survey) from the record. Otherwise, we may have no information on some items for a unit (partial non-response in a survey). Missing data may arise from other sources like damage to the records. And it must be noted that application of some techniques and some software for data analysis requires complete data sets.

Among different approaches to deal with missing data, the following are well-known.

1. Complete case analysis, which discards all incomplete cases (units in respect of which some of the observations are missing). While being simple, this method may be inefficient if the percentage of incomplete cases is large. There could be selectivity bias if selection is heterogeneous with respect to the co-variates.

2. Available case analysis, which differs from complete case analysis in the sense that here we make use of all observations which are complete in respect of the current step in analysis.

3. Imputation of the missing values, where the missing values are filled in by good guesses or correlation-based predicted values.

Three kinds of imputation can be distinguished: deductive imputations, deterministic imputations and stochastic imputations. Deductive imputations are imputations where the imputed value is deduced from known information. For example, the age of a person can be deduced from the given date of birth, and the total income can be found by adding the different income components. In a panel survey the year of birth of a person is constant, so if this variable is missing in a wave, the score on this variable from another wave could be copied to the missing value. Though it is a correct and often applied imputation technique, from a methodological point of view deductive imputation is less interesting than the other two kinds of imputation. Sometimes deductive imputation is considered part of the editing process.

A deterministic or stochastic imputation can be used if the correct value cannot be deduced. Then a prediction is made for the missing value. A stochastic imputation requires the generation of random numbers whereas a deterministic imputation does not. Hot-deck imputations can be either deterministic or stochastic imputation methods. If one record is drawn randomly from a group of records or if records are sorted randomly the hot-deck method is a stochastic method, otherwise the hot deck is a deterministic method.

An example of a deterministic imputation is the imputation of the mean of the known values of some variable for the missing values on that same variable. More accurate results can be obtained if groups of similar records are created and group means are imputed instead of the overall mean. It is a prerequisite that the records are classified into homogeneous groups. The idea behind imputing a group mean is that the unknown true value is better approximated by the group mean than by the overall mean. There is no guarantee that this is always true, but it is likely that a better prediction is found in most cases.

Even more accurate than mean imputation is imputing a regression result. This technique can be seen as a generalization of group mean imputation. First a regression is performed of the variable that has to be imputed on some explanatory variables. Only the records without item non-response on the variables in the regression are used to estimate the parameters. By means of the regression results, values are predicted for the missing values. Finally, these predicted values are imputed for the missing values on that variable. Unfortunately, regression imputation can only be used for missing values on continuous variables. Another disadvantage is that invalid scores (e.g. an age of 170 years) may be imputed with this imputation technique. Thus after a new round of editing we will again have item non-response.

An imputation method that uses information of another data set is the so-called cold-deck imputation. This method is especially appropriate for the imputation of panel surveys. In a panel survey, persons are interviewed several times. Values of another wave of the panel survey may be imputed for some of the missing values. This method may be improved as is demonstrated in the following example. Suppose that the income of a person in the last wave is known, but that there is item non-response on income in the current wave. Suppose further that it is known that the rise in the mean income between the last wave and the current wave is three per cent. If we let the income of the last wave grow by three per cent and impute the result for the income in the current wave, we have an improved prediction for the missing income. Further improvements may be obtained by taking into account the age of the person or the income rise in the economic sector where the person is working.

A disadvantage of deterministic imputations is that the distribution of the variable that has to be imputed generally becomes too peaked and thus the variance is underestimated. This distortion of the distribution is caused by the imputation of the best prediction at the record level. For advanced analysis this is unwanted and a worse imputation at the record level may be preferred so that the distribution of the variable that has to be imputed is less distorted. A more complete discussion about the shortcomings of deterministic imputations can be found in Kalton (1983). To preserve the variances and covariances of the imputed data, a random component is added to the deterministic imputation techniques to get its stochastic counterparts. A common way to introduce a random component in the imputation technique is to add a residual to the predicted value of the deterministic imputation. This residual can be selected from a normal distribution with expectation zero and variance σ^2. This σ^2 is equal to the residual variance of the regression of the variable that has to be imputed on some explanatory variables. The disadvantage of this method is that even if the original deterministic imputed value was feasible, the stochastic counterpart need not be. After adding the residual to the deterministic imputation, unfeasible values could namely result. Another way is to implicitly introduce a random component by imputing the value of a randomly selected person. This last method is called the random hot-deck overall method.

In the cold-deck method information from a record in another data set is used. For example the score on a variable of a similar record in the other data set is imputed for the missing value. If we compare the cold-deck method with the hot-deck method in the case of a panel survey we can say the following. In the cold-deck method the value of a variable of the same person in another wave is imputed, while in the hot-deck method the value of that variable of another person in the same wave is imputed. The record from which the imputed value is copied is called the donor record. In the hot-deck method, records from the current data set are normally used to provide imputed values for records with missing values. Basically, a search in the data set is made to find a donor record that matches the record with the missing value. The value from the donor record is then imputed for the missing value.

Matching the incomplete record with other records operates through the use of covariates, which can be defined by the user or can be found automatically. Instead of the word covariate, the term matching key is also used. Henceforth, we will for convenience use the word covariate. Records match if they have the same values on the covariates. For the covariates one can think of gender, age-group, the region where the person lives and the economic sector in which the person works. When more than one matching record has been found, based on some set of covariates, it is possible to decrease the number of matching records by adding another covariate. The idea behind using the covariates is that if two records have the same scores on the covariates, there is a high probability that the scores on the variable that has to be imputed are (almost) equal.

In the case of the sequential hot-deck method, the last matching record before the record with the missing value on the variable that has to be imputed is used as the donor record. This method is in principle a deterministic method but can be changed into a stochastic one by adding a residual as described in chapter 4. Changing the deterministic sequential hot-deck method into a stochastic imputation method can also be done by sorting the donor records randomly before the imputation takes place. This way of using the sequential hot-deck method is described in chapter 4.

Another possibility to obtain a donor record is to select one record at random from the matched records. This method is called the random hot-deck within-classes method. If there is only one class (no covariates) we have the random hot-deck overall method. If all records in the selection have the same value for the variable to be imputed, then we impute this value for the missing value; this is sometimes called an exact match. Mostly several records with different values for the variable to be imputed are available to serve as donor record for the imputation and then we draw randomly one of these records as the donor record; this is sometimes called a random match. In the exceptional case of a special order in the data set the sequential hot-deck method is sometimes preferred; normally the random hot-deck within-classes method is the best choice. Extensive reviews on the hot-deck method and other imputation methods can be found in e.g. Kalton (1983), Kalton & Kasprzyk (1986), Little & Rubin (1987) and Allen (1990).

The random hot-deck technique can also be used to impute several variables simultaneously. One of the variables that has to be imputed is then considered the main variable and imputed with e.g. the random hot-deck method. The same donor record is used for missing values on the other variables. This technique is called record matching. Instead of record matching the term 'common donor rule' is also used. Henceforth, we will for convenience use the term record matching. This technique is especially useful when some variables are strongly related and there are persons with missing values on several of these variables. Using record matching for the imputation of records where a group of variables is missing prevents inconsistency of imputations. Clearly, in this case only records where all the variables in this group have valid scores may serve as donor records. In the examples in chapter 4 it is considered important to preserve covariances and therefore the technique of record matching is often applied.

It is clear that if no sufficient data editing takes place, some inconsistencies in the data remain before the imputation process starts. Then inconsistent records may serve as donor records and thus the number of inconsistencies may rise. If we do not want to face this problem, we either have to edit more before the imputation process starts or we have to exclude the possibility that inconsistent records will serve as donor records. It can be concluded that applying record matching in the manner explained above will not introduce inconsistencies within the group of variables that is imputed. However, already existing inconsistencies in the data set may of course be transmitted if records with such inconsistencies serve as donor record.

A continuous covariate has to be divided into categories before the hot-deck imputation techniques can be used. The number of categories should be large enough to obtain a good imputation, but not too large because then the probability increases that no matching records can be found. Some methods do not require the continuous variables to be divided into categories first. One of these is regression imputation. A statistically more interesting method that excludes the possibility of imputing invalid scores is the method of predictive mean matching (Little, 1988a,b). This method is an interesting possibility for imputations with continuous covariates. The reason to apply predictive mean matching instead of a hot-deck method is that when using a hot-deck method, information is lost due to the temporary categorization of covariates.

If the method of predictive mean matching is applied, first a regression is performed of the variable that has to be imputed on some covariates. Again, only the records without item non-response on the variables in the regression are used to estimate the parameters. By means of the regression results, values are predicted for the missing values. For every missing value there is a search for the record with the nearest predicted value. Finally, the original value of this donor record is imputed for the missing value. As original values are imputed, all imputed scores are feasible values. The method of predictive mean matching is essentially a deterministic method, since, conditional on the regression model, the imputed values are always fixed by the values of the 'nearest neighbours'.

A stochastic counterpart could be created 'in the normal way' by adding a random component. Another way of making the method of predictive mean matching a stochastic method is the following. Instead of searching for the record with the nearest predicted value, one could also search for a group of records with predicted values close to the predicted value for the missing value. Then one could randomly draw one of these records as the donor record for the imputation, just as in the case of the random hot-deck within-classes method.

Besides the usual imputation techniques where one value is imputed for one missing value, the technique of multiple imputation can also be applied where several values are imputed for every missing value. The variation among all these values reflects the uncertainty under one model for non-response and across several models (Rubin, 1996). Each set of imputations is used to create a complete data set, each of which is to analysed as if there were no missing values. It can be shown that multiple imputation yields better inferences (Little and Rubin, 1989). More information about multiple imputation can be found in Rubin (1987). The merits of multiple imputation are discussed in Rubin (1996), Fay (1996) and Rao (1996) in the June 1996 issue of *JASA* and are followed by three interesting comments and rejoinders from the three authors.

Binder (1996) writes in his comment on the *JASA* 1996 articles of Rubin, Fay and Rao that one has to focus on the ultimate goals and purpose of the estimation process and on the assumptions that the data producer is writing to make about the response and data models. This implies that if the only interest is to estimate the mean or total of a given variable, and it is believed that the variable is independent of all other variables collected and the response mechanism is ignorable, then the imputation of the mean is a good solution. This imputation is best in the sense that no imputation variance is added. Although this is an extreme case, it is not too far removed from a large number of real cases. If the conditions are such that deterministic imputations will lead to biased results, then it is necessary to use a stochastic imputation. Stochastic imputations are assumed to yield (nearly) unbiased estimates. If the imputation variance is large, multiple imputation will be a good alternative to single-value stochastic imputation. However, the improved quality of the inferences and the cost of the extra computing must be weighted against one another. For some survey organizations limitations on resources are severe enough to adopt single imputation-and-variance-estimation methods.

The acceptance of a new procedure depends heavily on the extent to which the new procedure can be integrated efficiently with current valid survey practice. In the coming years more and more software with possibilities for multiple imputation can be expected to become available. With this software it must be easily possible to generate multiple imputations and to analyse multiply imputed data. Multiple imputation thus does not create information, but represents observed information so that it can be analysed using complete-data tools (Barnard, Rubin and Zanutto, 1997). Barnard, Rubin and Zanutto claim that for modest percentages of ignorable non-response multiple imputation with only a small number of replications yields consistent estimates and valid inferences. However, if the percentage of non-response is larger, the number of imputation replications has to grow to yield valid inferences. This may cause trouble to the survey organization, especially if the data set to impute is large.

It seems feasible that the problems with multiple imputation can be solved in the future so that this promising technique will then be applied more often in official statistics.

It is not easy to get an idea of the imputation quality because the true values are usually not known. One may compare means before and after imputation. External validations are seldom possible because of different definitions and target populations of other sources.

Little and Rubin (1987) classify the missing data mechanism according to the probability of response. If Y is the target variable and X is the covariate we can distinguish three cases:

1. the probability of response is independent of X and Y;
2. the probability of response depends on X but not on Y;
3. the probability of response depends on Y and possibly X as well.

If case 1 applies, we say that the data are missing completely at random (MCAR). In this case, the observed values of Y form a random sample of the values of Y. If case 2 applies, the data are missing at random (MAR). In this case, the observed values of Y form a random sample of the values of Y within classes defined by values of X. The missing data mechanism in case 3 is non-ignorable. In simulation studies ignorable non-response is usually assumed. However, in the practice of statistical offices non-ignorable non-response is a common phenomenon. Therefore, in this chapter imputation methods are tested stringently by introducing non-ignorable non-response in the data sets.

As the hot-deck method described above looks attractive, it is useful to study this method in more detail before applying it in the statistical production process. Simulation experiments may be conducted to study the performance of the hot-deck method. The experiments are designed to closely resemble empirical situations. The objective is to find out in a pseudo-empirical setting whether the hot-deck method yields less-biased results than methods that only use the information available in a variable with missing values. Thus we studied whether the hot-deck method is better than the available case method. In the second simulation experiment the effect of the covariates is also studied. Thus in the second experiment we also compare the random hot-deck within-classes method with the random hot-deck overall method.

8.8 Estimating Process Capability

On many occasions we come across several indices to measure the performance of a process or a procedure to meet certain contextual requirements. Thus to judge the performance of a manufacturing (production) process in relation to prescribed specifications for a quality characteristic X, we find many ways to define the process capability index, particularly when we deal with a measurable quality characteristic. Sometimes, a distinction is made between process potential index and process capability index. Given the same data on X, we get different ideas about process capability.

An estimate of the process capability index, say C_p for a process (of the quality characteristic X under study) symmetrically located in relation to the specification limits L (lower) and U (upper) or C_{pk} for a process not centred at $(U + L) / 2$, based on past and/or current observations on X is required to work out the capability of the process to meet the specification requirements and to assess the need, if any, for an adjustment in the process to change the process mean and/or the process variability. However, for a recently set (or re-set) process, it may be worth while to make use of a current sample and to consider the prediction interval for a future observation from the process and to compare the length of this interval with the specification range $R = U - L$. Of course, we have to speak of some specified content like a 100β % prediction interval with a high value of β like 0.995 or 0.990 or 0.950.

A prediction interval $[L_1 (X), L_2 (X)]$ for a function G (Y) of the random vector $Y = (Y_1, Y_2, ... Y_k)$ constructed on the basis of current sample observations $X = (X_1, X_2, ...X_n)$ is such that

$$[L_1 (X) \le G (Y) \le L_2 (X)] = \beta \ 0 < \beta < 1$$

where β is a pre-assigned constant. Here Y corresponds to a future sample. This is known as 100β % prediction interval for G (Y). In the simplest case, we can consider G (Y) as just Y or we could take it as the mean of, say, m future observations on Y.

Consider the following measurements of the diameter (mm) X of 25 piston rings for an automobile engine. The target process mean is 74.0 and the specification range is 0.12. It can be verified that the distribution of X is reasonably normal.

74.030	73.995	73.998	74.002	73.992
74.009	73.995	73.985	74.008	73.998
73.994	74.004	73.983	74.006	74.012
74.000	73.994	74.006	73.984	74.000
73.998	74.004	74.010	74.015	73.982

The sample mean comes out as 74.000 and the standard deviation s = 0.019.
Going by the classic definition of CPI, we get $C_p = (U - L) / 6 s = 012 / 0.114 = 1.06$.

If we decide to take the length of the 100β % prediction interval based on the current sample of 25 observations, we have by the frequentist approach

$$C_{PI}^F = [U - L] / [2 \sqrt{(1 + 1 / n)} \, s \, t_{(1-\beta)/2, \, n-1}]$$

where we take the value of t-distribution with $n - 1$ degrees of freedom exceeded with probability $(1 - \beta) / 2$. In this case the value of C_{PI}^F works out as 1.8.

Extending the computation to the Bayesian framework, we assume a quasi-prior for the process mean μ and process SD σ with the joint probability density function as $h_1 (\mu, \sigma) = \text{const. } 1 / \sigma^c$. Obviously, $h_1 (\mu, \sigma)$ is a proper prior if $c > 1$.

Values of the PCI index using quasi-prior C_{PI}^Q are shown below for different c values.

c	C_{PI}^Q
1.0	1.8263
5.0	1.9970
10.0	2.2153

It is found that as the value of c increases, process capability using a prediction interval also increases. In practice, it hardly makes sense, except to tell us that if the prior density is smaller, the estimated process capability gets higher.

We can as well use the conjugate prior for μ and σ with p.d.f.

$$h_2 (\mu, \sigma) = g_1 (\mu / \sigma) g_2 (\sigma)$$

A natural choice for the distribution of μ given σ is normal with the p.d.f.

$$g_1 (\mu / \sigma) = \exp [- (\mu_0 - \mu)^2 / 2 \sigma_0^2] / [2 \pi \sigma_0^2 \sigma^2]^{\frac{1}{2}}$$

$$- \infty < \mu, \mu_0 < \infty, \sigma > 0 \text{ with } E (\mu / \sigma) = \mu_0 \text{ and } Var (\mu / \sigma) = \sigma^2 \sigma_0^2$$

The prior for σ is taken as inverse gamma with the p.d.f.

$$g_2 (\sigma) = \frac{\alpha^{(p-1)/2} e^{-\alpha/2\sigma_0^2}}{\sigma^p \, 2^{(p-3)/2} \, \Gamma [(p-1)/2]} \qquad \alpha > 0, p > 1$$

Here $E (\sigma) = \sqrt{(\alpha / 2)} \, \Gamma (p/2 - 1) / \Gamma [(p-1) / 2]$ and

$$Var (\sigma) = \alpha / (p - 3) [1 - 2 / (p - 1) \Gamma^2 (p / 2 - 1) / \Gamma^2 (p - 3 / 2)]$$

By choosing different combinations of p and α values, we get quite different values of process capability, as given in the Table 8.6. In fact, the wide range of PCI values with different hyper-parameter values as given in the table is almost baffling to a user.

As expected, PCI increases with increasing p / α. As σ_0 increases, PCI decreases. This example clearly brings out the fact that with different choices of the priors of process mean and variance, we may face a wide divergence among the PCI values. Hence the need to

TABLE 8.6

Value of PCI Using Conjugate Priors

	p	$\alpha = 0.001$	$\alpha = 0.01$	$\alpha = 0.05$
$\sigma_0 = 0.1$	5	3.9741	2.2380	1.1146
	10	4.3859	2.2470	1.2300
$\sigma_0 = 0.5$	5	1.9142	1.0783	0.5368
	10	2.1125	1.1900	0.5925
$\sigma_0 = 1.0$	5	1.8125	1.0210	0.5083
	10	2.0009	1.1268	0.5610

understand the role of a prior distribution and to take any PCI value under the Bayesian set-up with due caution.

8.9 Estimation of EOQ

Imagine a research worker who wants to estimate the optimal order quantity based on a sample of demands along with given costs c_1 and c_2 of shortage and excess (per unit short / in excess per unit of time) respectively and using the traditional mean-minimizing solution to a static inventory problem (often referred to as the classic newspaper boy problem). This is quite an interesting and useful exercise, in view of the fact that the optimal order quantity assuming a completely specified demand distribution will have little value in reality. And it may be quite possible to have a sample of demands which can be used to estimate the optimal order quantity. It can be easily shown that the optimal order quantity works out as EOQ $q_0 = F^{-1} [c_1 / (c_1 + c_2)]$, with F as the demand distribution function.

Since the EOQ happens to be the quantile of order $c_1 / (c_1 + c_2)$ of the demand distribution, we can use a bootstrap estimate quite generally. In fact, bootstrapping has been quite often used to estimate quantiles from an unknown distribution.

In the case of demand following a one-parameter exponential distribution with probability density function as $f(x: \lambda) = \lambda \exp(x)$, the optimal order quantity is simply given as $q_0 = -1/\lambda \ln k$ where $k = (c_1 + c_2) / c_1$. A non-parametric estimate of EOQ will be the sample quantile of order $1/k$, which will be asymptotically unbiased. A simple parametric estimate will be $-x$ bar $\ln k$, which also will be unbiased. To accommodate possible variations in the parameter we can introduce the natural conjugate prior, viz. gamma with shape parameter p and scale parameter α with the prior expectation p / α. Assuming a squared error loss, the Bayes estimate of EOQ comes out as

$$Q_0 = -k_1 - \text{xbar } k \text{ where } k_1 = \alpha / (n + p) \text{ and } k = n / (n + p) \ln c_2 / (c_1 + c_2)$$

with a demand sample of size n. If we want to replace the unknown hyper-parameters p and α by their sample estimates, viz. $p' = xx$ bar and $\alpha' = x$ bar $/ s^2$, then the empirical Bayes estimate of EOQ is found as

$$q_{0 \text{ (EB)}} = -(\alpha^{\wedge} + \Sigma x) / (n + p^{\wedge}) \ln c_2 / (c_1 + c_2)$$

However, if we use the lines loss function given by $L(\delta) = b\{e^{a\delta} - a\delta - 1\}$ $b > 0$, $a > 0$ and $\delta = \lambda^* - \lambda$, λ^* being the Bayes estimate, the estimated EOQ in this case is $-1/\lambda^* \ln c_2 / (c_1 + c_2)$.

It may be interesting to illustrate the alternative estimates making the situations comparable by having the same mean demand.

Let $\lambda = 1/20$, $p = 30$, $\alpha = 1.5$, $n = 10$, $\Sigma x = 380$ c_1, $c_2 = 0.5$, $a = b = 40/3$. Then

Solution	EOQ	Minimized Cost
Classic	2.703	5.466
Bayes with squared error loss	1.392	2.784
Bayes with lines loss	1.399	2.798

It thus comes out that both the estimated EOQ and the associated cost are smaller with the Bayes approach than in in the classic approach.

In many situations, we can assume demand to follow a beta distribution with parameters m and n over the range (a, b). Without losing any generality, we can take the range of demand as (0, 1) to have the demand p.d.f. as

$$f(x; m, n) = 1/B(m, n) \, x^{m-1}(1-x)^{n-1} \, m, n > 1$$

Assume, for the sake of simplicity, $c_1 = c_2 = 1$. The expected cost associated with an order quantity q is given by

$$E[C(q)] =$$

Since an incomplete beta integral will not permit an analytic expression for finding the optimal value of q even when the parameters m and n are known, we first estimate the parameters by either equating the sample and variance to the corresponding functions of m and n or by equating the sample mean and the sample mode to the parameters and put the estimates in the expression for $E[C(q)]$ to get the estimated expected total cost associated with an order quantity q. Subsequently, we try a power series approximation to the incomplete Beta integral involved in the expression for the expected total cost to get a differentiable function.

Now the estimated expected total cost can be differentiated w.r.t. q to yield the optimal order quantity and the minimum cost associated with it.

8.10 Comparison among Alternatives Using Multiple Criteria

8.10.1 Some Points of Concern

An important task sometimes faced in research is to compare among several alternatives based on more than one criterion. The idea is either to choose the 'best' alternative or a handful of 'better' alternatives including the 'best' along with a few others close to the 'best' to attach priority indices to all the alternatives. The alternatives could be different brands of the same product, different designs for an industrial experiment, different software to execute the same complex computation, different candidates to discharge a given responsibility or even different criteria to be adopted for the purpose of comparison. The alternatives could be just several different situations or entities which should be ordered according to some scale of preference. For example, one may compare several different cities for which we may want to find out the quality of the environment influencing liveability.

Selection or prioritization is a decision and here one is involved in multi-criteria decision-making (MCDM). Quite often a criterion for comparison may include a number of attributes possessed by the alternatives. For example, the criterion of 'dependability' of an individual for some task may imply competence, honesty, timeliness etc. as different attributes; the criterion 'cost' for comparing alternative brands may include components of life-cycle cost like cost of acquisition, of usage (in terms of energy required), likely cost of maintenance, re-sale value etc. For software, we may consider user-friendliness, accuracy, re-usability, inter-operability, etc. as attributes to be taken into account in any comparison. This is why such decision-making problems are referred to as multi-attribute decision-making (MADM). We will subsequently use the terms 'criterion' and 'attribute' as synonymous. It may incidentally be pointed out that 'attributes' in this connection could be discrete or continuous 'variables' in some cases.

Ranking and rating are the usual methods for comparison. Ranking among a few and more or less easily differentiable alternatives according to a single criterion or attribute results in a unique rank assigned to any one alternative. In other situations, specially when the alternatives are not obviously different one from the other, ranking will result in ties or equal ranks assigned to a few individuals assessed as equivalent. In the case of a large number of alternatives, beyond a certain small rank, alternatives with high ranks may not reflect the true situation. And all this for a single criterion. For the time being we assume a single judge to assign ranks. If we have to consider multiple attributes (with a single judge) or with multiple judges (taking care of a single attribute), we encounter the problem of possible disagreement among the multiple rankings. In either case, it will not be prudent to compute the average ranks for our purpose without checking the extent of agreement or concordance among the rankings. We can use Kendall's coefficient of concordance and test for its significance. A high degree of concordance would justify averaging of ranks and the average ranks can help us in reaching a decision.

Realizing the fact that rank differences do not reflect equal differences in the level of an attribute over different parts of the ranked set, one may think of rating the candidate objects or individuals, assigning some scores to each. Both ranking and rating are based on subjective assessment. To distinguish one member in a pair as better or worse that the other member can most often be done much more objectively than ranking the whole set of candidates. Thus, we can have recourse to pair-wise comparisons and Thurstone's product-scaling method based on the assumption of an underlying normal distribution of the trait or attribute to get scale values for each of the candidates. Although, historically, product scaling involved several judges and a single attribute for each product, the method can easily be applied to deal with multiple attributes and a single judge. Each pair of products is presented to the judge to assess one as better than the other. And this pair-wise comparison is to be done for each attribute. We develop a matrix $((p_{ij}))$ where p_{ij} denotes the proportion of attributes in relation to which product I is judged as better than product j. This is equated to the area beyond $\mu_i - \mu_j$, where μ_i is the mean value of the underlying trait for product i and similarly for product j. Equating the mean scale value to be 0, we get scale values for all the products.

A widely used method for prioritization has been the technique for order preferences based on similarity to the ideal solution (TOPSIS) introduced by Hwang and Yoon

(1981). Of course, the analytic hierarchy process (AHP) and the analytic network process (ANP) introduced by Saaty have been well-established methods for multi-attribute decision-making. Also occasionally used is operational competitiveness rating analysis (OCRA) in situations where inputs for each alternative along with the corresponding outputs are noted for each case (maybe a decision-making unit) and usually a single criterion like productivity or technical efficiency is involved. In fact, a widely used method for comparing the decision-making units in such situations is data envelopment analysis and its many ramifications. We present below brief accounts of some of these methods.

8.10.2 Analytic Hierarchy Process

AHP generates a weight for each evaluation criterion based on the decision-maker's pair-wise comparison among the criteria to get over the problem that no alternative may be found the 'best' according to all the criteria. A higher weight implies a greater importance for the criterion in making overall prioritization. Next, for a fixed criterion, AHP assigns a score to each alternative according to the decision-maker's pair-wise comparison among the alternatives based on that criterion. Finally, AHP combines the scores for the alternatives and the criteria weights. A final prioritization of the alternatives is in terms of the weighted total of score for each. The pair-wise comparisons which form the bedrock of AHP are somewhat analogous to Thurstone's product scaling, though the latter does involve several judges or decision-makers stating their preferences for one product or alternatives compared to the a second where the two are presented as a pair. Moreover, Thurstone's method does not involve multiple criteria.

AHP can be implemented in three simple computational steps viz.

(a) computing the weight of each criterion, yielding a vector

(b) computing the matrix of scores for each alternative in respect of each criterion

(c) ranking the alternatives based on the weighted total scores.

With k criteria, pair-wise comparison by the decision-maker yields the $k \times k$ matrix A with elements a_{ij} where $a_{ij} > 1$ if criterion i is regarded as more important than criterion j, $a_{ij} < 1$ otherwise. Obviously, $a_{ii} = 1$ for all i. If both the criteria are equally important, then also $a_{ij} = 1$. The entries a_{kj} and a_{jk} satisfy the constraint $a_{jk}. a_{kj} = 1$.

The relative importance of one criterion relative to the other, when presented in a pair, is measured on a scale from 1 to 9 as follows

Value of a_{jk}	Interpretation
1	criteria j and k are equally important
3	criterion j is slightly more important than k
5	criterion j is more important than criterion k
7	criterion j s strongly more important than k
9	criterion j is absolutely more important than k

It may be incidentally mentioned that the strength of relation between a product feature and a technical parameter is generally expressed on a scale of 1, 3, 6 and 9 in constructing the house of quality in quality function deployment.

We now normalize the entries in this matrix to make the column sums equal one and we get elements $b_{ij} = a_{ij} / \Sigma_i b_{ij}$. Finally, these normalized entries are averaged over rows to yield the criterion weights as

$$w_i = \Sigma_j b_{ij} / k \text{ for } i = 1, 2, \dots k.$$

Consider an example where the quality of water in a river in three different ecological regions is to be arranged in a hierarchy based on four important quality parameters, viz. dissolved oxygen, total dissolved solids, total coliform count and the ratio between bio-chemical oxygen demand and chemical demand. First, we present to an expert or the decision-maker these quality criteria in pairs and we obtain the following matrix

	1	3	5	1/3	It can be seen that $a_{jk}. a_{kj} = 1$ for all pairs (j, k).
A =	1/3	1	1/7	1/5	Criterion 4 has been regarded as the most important
	1/5	7	1	1/3	and criterion 2 as the least important.
	3	5	3	1	

The corresponding normalized matrix with column sums equalling unity works out as

0.2205	0.1875	0.5469	0.1786
0.0735	0.0625	0.0156	0.0710
0.0441	0.4375	0.1094	0.1786
0.6617	0.3125	0.3281	0.5357

The row averages yield the vector of criteria weights as $\mathbf{w}' = (0.2834, 0.0647, 0.1924, 0.4595)$. To check the consistency of paired comparisons yielding the criterion weights, we calculate the ratio of elements in the vector $\mathbf{Aw}$ to the corresponding elements in the weight vector $\mathbf{w}$ and these come out as 5.61, 4.22, 4.37 and 4.78. The consistency index CI is obtained as $CI = (x - m) / (m - 1)$, where x is the average of the ratios just obtained. Thus CI works out as 0.25. To interpret this value, we note the value of the random index, i.e. the consistency index when the entries in A are completely random. The value of this RI for m = 4 is given as 0.90 so that CI / RI = 0.28, which indicates slight inconsistency.

Given the decision matrix with values of the four parameters after some appropriate transformation appearing in the columns and rows representing the regions as indicated below, we find the weighted score for each of the three regions.

Region	DO$_2$ (%)	TDS	TCC	BOD / COD
1	1.8	14	21	2.0
2	2,2	16	20	1.8
3	2.0	15	17	2.1.

These come out as 1.9237, 1.9389, 1.9559 respectively. Since these figures relate to aspects of water pollution, the best region should be region 1 – a fact not apparent from values in that row compared to values in the two other rows. One may argue that dissolved oxygen is really not an aspect of pollution, one could consider values in column 1 with a negative sign, to yield weighted totals 0.9035, 0.6919 and 0.8223 respectively, and this implies region 2 is the best. In any case, what must be remembered is that the weights were obtained without taking account of the decision matrix.

8.10.3 Data Envelopment Analysis

Data envelopment analysis (DEA) is a relatively simple non-parametric method to compare the efficiencies of multi-input, multi-output decision-making units (DMUs) based on observed data. Such units are engaged in producing similar outputs using similar inputs, most often working within the same corporate organization but each enjoying some amount of authority to decide on the deployment of the resource inputs. Examples could be branches of a bank, schools working under the same board or council, hospitals under the same health department, service units of a municipal corporation, and the like. Sometimes, the DMUs could be different organizations in the same business. Efficiency (sometimes referred to as technical efficiency) is defined as the ratio between the weighted total of outputs produced and the weighted total of inputs used, where weights are to be endogenously generated from the data and not exogenously assigned by experts. In fact, these weights are determined by repeated application of linear programming in DEA. DEA is preferred to other comparable forms in econometric analysis which assume some parametric model linking outputs as functions of inputs.

With n DMUs consuming varying amounts of m different inputs to turn out s different outputs, let DMU j consume amount x_{ij} of inputs i = 1, 2, ... m to produce an amount y_{rj} of outputs r = 1, 2, ... s. Let U and v be the vectors of weights to be associated with the input vector X_j and the output vector Y_j for any DMU j. Then the original input-oriented DEA model to find the weights u and v and therefrom the efficiency for each DMU as proposed by Charnes, Cooper and Rhodes (as the CCR model) can be stated as

$$\text{Maximize } w_0 = v' Y_0 \text{ subject to } u' X_0 = 1, v' Y_0 - u' X_0 \leq 0, u, v \geq \varepsilon$$

The problem is solved n times with $\{X_0, Y_0\} = (X_j, Y_j)$ j = 1, 2, ... n. Values of w partition the set of DMUs into efficient DMUs with w = 1 and inefficient ones with w < 1, with the efficiency frontier corresponding to w = 1. The output-oriented model, which is just the dual of the LP problem stated earlier, was proposed by Banker, Charnes and Cooper as the BCC model.

Other DEA models include multiplicative models that allow a piecewise log-linear or piecewise Cobb-Douglas envelopment. If we omit the contraction factor φ and focus on maximizing the sum of the slacks in the BCC model, we get another variant of DEA. If we have some prior knowledge about the weights or factors, we can incorporate such knowledge in the model by, for example, putting lower and/or upper bounds on their values, resulting in less flexible models compared to the original CCR or BCC models where no restrictions on the factors or weights are introduced (Dyson and Thanassoulis, 1988). Banker and Morey (1986) had shown how to incorporate categorical variables in DEA. Sahoo (2015) considered a model with variable returns to scale as opposed to the constant returns assumed in the classic models. There has been some contribution to the dynamic DEA model. Johnson and Kuosmanen (2012) consider the use of infinitely many DMUs. Tone (2000) used a limited number of additional linear programming problems to be solved in respect of the inefficient DMUs to work out the possible improvement in their efficiency by significantly reducing their consumption of inputs or production of outputs. Applications of DEA models have been wide-ranging, from banking to school education, from health care to transportation, from corporate failure analysis to bankruptcy prediction problems, and many others.

One limitation of DEA is that it does not provide a basis for comparing efficient DMUs among themselves. It is quite possible for several DMUs to have unit efficiency score in the classic DEA models. For this purpose, super-efficiency models based on slacks have been proposed. Also a cross-efficiency matrix can be constructed and the column averages can distinguish among the efficient units.

If we recognize random variations in inputs and/or outputs because of errors of measurement or of sampling or other chance factors, we have to formulate and solve a stochastic DEA problem. One approach will be to look upon DEA as a problem in chance-constrained programming problem (Olesen and Petersen, 2003) with both constraints and the objective function as random and to come up with chance-constrained efficiency index. The problem thus formulated is a non-convex programming problem and to avoid computational complexity, upper and lower bounds for the efficiency measures were suggested. Sengupta (1987) also studied the chance-constrained case for single-output units. Statistical analysis done by Banker (1993) showed that DEA provides the maximum likelihood estimate of the efficiency frontier. A bootstrapping method is proposed to estimate the frontier. Morita and Seiford (1999) considered efficiency as a random variable and discussed the expected efficiency score, quantiles of the efficiency score and probability of the score exceeding a threshold value as possible efficiency measures and introduced the concepts of reliability and robustness of the efficiency result. If an efficient DMU remains efficient for large deviations in inputs and outputs from their assumed values, such a DMU has a high reliability. Otherwise if an efficient DMU turns out to be inefficient for small changes in data, such a DMU is not reliable.

8.10.4 TOPSIS

Among all MADM methods, the technique for order preference by similarity to ideal solution (TOPSIS) is quite an effective one in terms of its intuitive appeal, nice construction and computational simplicity. TOPSIS uses the intuitive principle that the selected alternative should have the shortest distance (Euclidean) from the best solution and the farthest distance from the worst solution. A solution is a point with coordinates as the attribute values/levels. In a higher the better situation (originally or after an inverse transformation of an attribute), the best solution contains the maximum attainable values of the attributes and the minimum values constitute the worst solution.

In one MADM approach, the alternative which has the shortest Euclidean distance from the best solution is selected. However, a solution that has the shortest distance from the best may not be farthest from the worst. TOPSIS considers both these distances simultaneously. This distance may be denoted by σ (B, W). The decision criterion in TOPSIS is that the smaller the value of d (B, W) the more the alternative is preferred.

TOPSIS has been applied in various contexts. One important application has been in the area of quality function deployment. A slightly modified version called A- TOPSIS has been developed for comparing alternative algorithms.

Unlike in AHP, weights for the different decision criteria are determined from the actual data in TOPSIS and from aired comparisons among the criteria by some experts or by the decision-makers, without taking the data into account. In fact, the weight for a given criterion in TOPSIS is obtained using the entropy measure of the observed values corresponding to this category or using the coefficient of variation or some other principle.

To bring out differences between AHP and TOPSIS, in terms of procedure and results, we will apply the entropy principle for determining the criteria weights. We will consider the previous example. Let x_{ij} be the element in the (i, j)th cell of the decision matrix.

Step 1. Elements of the decision matrix are normalized to obtain probabilities estimated in terms of proportions a $p_{ij} = x_{ij} / \Sigma_i x_{ij}$ j = 1,2, ... n.

Step 2. Compute the entropy measure for each criterion j as $E_j = - \alpha \Sigma_I p_{ij} \ln p_{ij}$ for j = 1,2, ... n.

These measures for the example considered in sub-section 8.10.2 work out as

$$- E_1 = 0.4623, - E_2 = 1.0993, - E_3 = 1.0335 \text{ and } - E_4 = 0.4578$$

The degree of diversity is then calculated as $D_j = 1 - E_j$. The entropy weights W are then obtained as $W_j = Dj / \Sigma D_I$ for j = 1, 2, ... n. Let us compute these weights for the example to get

$$W_1 = 0.2073 \qquad W_2 = 0.2977 \qquad W_3 = 0.2883 \qquad W_4 = 0.2067$$

Step 3. Transform the decision matrix **X** to a normalized matrix R = $((r_{ij}))$ where $R_{ij} = X_{ij} / \sqrt{\Sigma x_{ij}^2}$ so that $\Sigma_{ij}^2 = 1$.

Step 4. Elements in the above matrix are weighted by the appropriate criteria weights to yield the following matrix: **V** = $((v_{ij}))$ where $v_{ij} = W_j r_{ij}$.

Step 5. Define the 'ideal positive' and 'ideal negative' solutions indicated by

$$V_j^+ = \min v_{ij}, \text{minimum over i} \quad \text{and } V_j^- = \max v_{ij} \text{maximum over iI for j1, 2} =,...n$$

Step 6. Compute the distance of entity I from the positive ideal and negative ideal as

$$D_I^+ =]\Sigma(v_{ij} - v_j)^2]\text{‰and } d_i^- =\left[\Sigma\left(v_{ij} - V_j^-\right)^2\right]\text{‰the sums being over j.}$$

Step 7. Compute the composite index for each entity to determine the closeness of an entity from the ideal solution in terms of $CI_I = d_i^+ / (d_i^+ - d_i^-)$ I = 1, 2, ... m. The final rank for entity k is obtained this way and the higher the rank the closer the entity is to the ideal solution.

8.10.5 OCRA

A new performance measurement procedure called operational competitiveness rating (OCRA) has been proposed (Parkan, 1994) essentially as a benchmarking tool to measure the performance efficiency of a production unit within a set of such units. Based on a nonparametric model, the OCRA procedure involves the solution of a linear programming problem to obtain an inefficiency index of each production unit relative to other units within the set considered. The index is computed on the basis of inputs consumed and outputs yielded by each unit. This non-parametric procedure is somewhat akin to data envelopment analysis (DEA). The procedure has been applied to banks and other service-providing units besides manufacturing and processing units. OCRA has been criticized also for the subjectively determined calibration points used as weights for inputs and outputs to come up with a simple ratio to be optimized. The complementary perspective of developing an index of relative inefficiency has also been applied in practice to costs of input resources and values of output goods and services.

Resources consumed by each of k production units PUs are put into m cost categories and values generated by output goods and services in N revenue categories. Then the resource consumption performance rating and the revenue generation performance rating for PU_I are obtained to subsequently yield the overall performance rating. Critics of OCRA point to the subjectivity of the calibration constants expected to reflect management's perceived priorities. A direct and possibly elaborate method for this purpose could be the analytic hierarchy process. Wang and Bakhai (2006) pointed out that in OCRA a cost category with large cost/revenue ratio is more important than a cost category with a small ratio, creating an illusion to management

8.11 Conjoint Analysis

Conjoint analysis has been a useful survey-based technique making use of statistical methods to find out quantitatively the preferences of customers for different attributes and their levels possessed by or to be designed into products or services. Although applied as a market research tool for designing new products or changing or repositioning existing products, evaluating the effects of price on purchase intent, and simulating market shares, conjoint analysis really attempts trade-offs among different attributes which cannot behave in the same direction in the actual design or development or purchase situation. In fact, conjoint analysis works out the utility or part-worth of each attribute in the assessment of the customers or potential customers. Such assessments may be in terms of ranks or grades or in terms of a dichotomy, viz. preferred or not. While purchasing a lap-top computer, one may consider attributes like the brand, the screen size, the battery life, the memory size and also the price. And possible specifications for each attribute define the levels of the attribute. The product can be regarded as a combination of attribute levels.

Subjects who are potential decision-makers provide data reflecting their preferences for hypothetical products which are defined by combinations of attribute levels like treatment combinations in a designed experiment. Any combination of attributes including the case of a single attribute is called a concept. A combination of levels of all the attributes is called a full-profile concept. All possible combinations need not be considered for the survey; attributes can be severally presented to the potential buyer to extract the preference for each attribute. Alternatively, attributes may be presented as combinations or conjoined. All possible levels of a preferred attribute may not be admissible and only those which are acceptable to the potential buyers may be included.

When all the attributes are nominal, the metric conjoint analysis is a simple main-effects ANOVA with some predefined output corresponding to the judgement of a decision-maker. With three attributes we can have a linear model

$$Y_{ijk} = \mu + \beta_{1i} + \beta_{2j} + \beta_{3k} + \varepsilon_{ijk} \text{ where}$$

$\Sigma \beta_{1i} = \Sigma \beta_{2j} = \Sigma \beta_{3k} = 0$. In this model, β_{1i} is the respondent's stated preference for a product with the ith level of the first attribute, and similarly for the other two attributes. The predicted utility of a product represented by the ith level of attribute 1, jth level of attribute 2 and kth level of attribute 3 can be obtained by estimating the coefficients β. In non-metric conjoint analysis we replace the response y_{ijk} by a monotone transform. The coefficient of determination R^2 in non-metric analysis is always larger. If we have a discrete choice to indicate preference, we use a non-linear logit multinomial model. The transformation finally used is often found iteratively.

In the case of two levels of each factor, computations become simpler. Utilities for the two levels of each such factor have the same magnitude with different signs. The importance of an attribute is the percentage utility range., computed as the range of utility for the attribute (which is double the utility for the better level) divided by the sum of all such ranges (across attributes).

Choice-based and adaptive conjoint analysis require full-profile concepts, the latter focusing on fewer concepts in terms of levels of features or attributes that are preferred by the buyers. Explicated conjoint analysis does not require full profiles, Instead preferences for each attribute level are sought from the respondents.

8.12 Comparison of Probability Distributions

A problem somewhat related to the one discussed in the previous section is the problem of comparing the observed distribution of a single variable or the joint distribution of several variables in one community or group with that in another. We can obviously have more than two groups to be compared. Thus we may be interested to compare the distribution of income or wealth or productive assets or per capita consumption expenditure or level of living etc. among several regions in a country or among different socio-economic groups. We take interest in comparing the mortality pattern or income inequality across countries.

Observed distributions of some variable like income or consumption expenditure or age at death etc. yielded by sample surveys can often be graduated to find the corresponding probability distributions. Usually, we compare the mean values and sometimes we also compare some measure of dispersion like the standard deviation. In the context of theoretical research we may have to compare two distributions, e.g. of costs associated with two different order quantities, and, in such cases, we simply compare the expectations to find out if one cost distribution dominates over the other, in which case we will prefer the second order quantity.

Considering, for the time being, two non-negative random variables X and Y with distribution functions $F(x)$ and $G(x)$ respectively and probability density functions $f(x)$ and $g(x)$, we state that Y dominates X in distribution or stochastically, if $G(x) \geq F(x)$ for all $x \geq 0$ and we denote this relation as $Y >_d X$.

If, in addition, $G(x) > F(x)$ for at least some x, then there is strict stochastic dominance. Integrating both sides over the entire range of x, we get $E(Y) \geq E(X)$, which implies expectation dominance of Y over X, denoted as $Y >_E X$. It may be pointed out that expectation dominance does not necessarily imply stochastic dominance though stochastic dominance does imply expectation dominance. In fact stochastic dominance means the right tail area beyond any value x under the distribution curve of Y is larger than or equal to the corresponding area under the distribution curve of X. Strict dominance requires the inequality > for at least one value of X.

Higher-order stochastic dominance is defined in terms of integrals of distribution functions. Thus X is smaller than Y in the second-order stochastic sense, denoted as $Y >_{d^2} X$, if $\int_x [1 - F(t)] dt < \int_x [1 - G(t)] dt$ for all x. Second-order stochastic dominance can be equivalently stated in terms of the relation $\int F^{-1}(s) ds \leq \int G^{-1}(s) ds$, integrals over

[0, t], where F^{-1} denotes the usual quantile of order s, i.e. $F^{-1}{}^{(t)} = \inf. \{x: F(x) \geq t\}, 0 \leq t \leq 1$. Similarly, one can define inverse stochastic dominance in terms of the quantiles, to write $X >_{I-SD} Y$ if $F_x^{-1}(t) \geq F_Y^{-1}(t)$ for all t in (0, t).

It can be shown that first- and second-order stochastic dominance are equivalent to first- and second-order inverse stochastic dominance respectively.

In the context of income distribution comparisons, first-order stochastic dominance implies that the head count ratio (of people below a given income level, e.g. the poverty line) is higher in the dominated distribution. Second-order stochastic dominance can be stated also in terms of the inequality $\int (x - t) f(t) dt \leq \int (x - t) g(t) dt$ over the range (0, x) for all x. Here, it is worthwhile to note that the integral represents the total income gap from the level x (e.g. total income gap of the poor).

Lower-order stochastic dominance implies higher-order stochastic dominance.

If **X** and **Y** are two p-dimensional random variables, then **X** is stochastically dominated by **Y** if $E[f(\mathbf{X})] \leq E[f(\mathbf{Y})]$ for all $\leq Pr(\mathbf{X} > x) \leq Pr(\mathbf{Y} > x)$ for all **x**. And **X** is smaller than **Y** in lower orthant order if $Pr(\mathbf{X} \leq x) \geq Pr(\mathbf{Y} \leq x)$ for all **x**.

X is smaller than Y in likelihood ratio order if the ratio g(t) / f(t) increases in t over the union of supports of X and Y, assuming the random variables to be continuous with probability density functions f and g respectively.

X is less than Y in the convex order if and only if for all convex u $E[u(X(] \leq E[U(Y)]$. Convex ordering is a special kind of variability ordering, narrowly captured by a comparison of variances in the distributions compared. Laplace transform order compares both the size and variability of two random variables. Laplace order is established in terms of an inequality similar to that in convex order where $u(x) = -\exp(-\alpha x)$ with α a positive real number.

In the context of reliability and survival analysis, we speak of failure (hazard) rate ordering and mean residual life ordering. As expected, these are in terms of inequalities involving failure rate of a random variable X being defined as $r(t) = f(t) / [1 - F(t)]$ and $\mu(t) = E[X \mid X \geq t] - t$ respectively

where f and F are the density and distribution functions of X respectively.

Another aspect of variability or dispersion is peakedness (not just measured by kurtosis) according to which partial orders among distributions may also be of interest in some context. Birnbaum (1948) defined the peakedness of a random variable X around a value a as $P_a(x) = Pr(\mid X - a \mid \leq x) x \geq 0$ for a distribution symmetric around a. The existence of moments is not needed and lack of symmetry can be easily incorporated. Going by this definition, distributions, random variables X and Y with distribution functions F and G are said to be ordered in respect of peakedness, denoted by $F \leq_{p(a,b)} Y$ if

$$Pr(\mid X - a) \leq x) \leq Pr(\mid Y - b \mid \leq b) \text{ for all } x \geq 0.$$

We can put a = b = 0 without any loss of generality. It can be easily shown that peakedness ordering does imply stochastic ordering of $\mid X \mid$ and $\mid Y \mid$ in the sense that

$$Pr(\mid X \mid \leq t) \leq Pr(\mid Y \mid \leq t) \text{ for all } t \geq 0.$$

Partial orders according to dispersion measured by variance and peakedness as defined above have been extended to multivariate cases. Sherman considered peakedness ordering between two random vectors **X** and **Y**. These partial orders have been applied to develop optimality criteria for the design of experiments (Sinha, 2002).

Estimation of probability distributions with a known partial order between them is quite useful in, say, estimating reliability from stress-strength considerations, given that strength as a random variable will have a distribution with some known order with the distribution of stress.

One can find applications of stochastic dominance concepts in the context of ROC. If t denotes the threshold value of the bio-marker or some other differentiating characteristic X and F and G denote the distributions of X in the normal and abnormal or affected or suspected population respectively, it is expected that values of X (if necessary, through some transformation) under G will be larger than those under F. This really implies that G stochastically dominates F so that $E(X / F) < E(X / G)$. We can think of bivariate ROC analogues by considering two bio-markers to discriminate between two groups and consider bivariate stochastic dominance.

8.13 Comparing Efficiencies of Alternative Estimation Procedures

We sometimes like to compare alternative procedures for estimating the same parameter in terms of their (asymptotic) variances and prefer some procedure(s) as better than some other(s). It is commonly claimed that an estimator based on inspection by variables is more efficient than the corresponding estimator based on inspection by attributes. We tend to forget that these two alternative procedures are not strictly comparable in terms of time, operational cost and other factors. Thus the reliability of a piece of equipment R to survive till a mission time t_0 can be estimated either from the number r of copies in a random sample of items/copies surviving up to t_0 or from times-to-failure $t_1, t_2, \ldots t_n$ on the assumption of a failure-time distribution. Though not universally tenable, the one-parameter exponential distribution is a simple and widely used model for this purpose. We assume the p.d.f. to be $f(t; \mu) = 1/\mu \, e^{-t/\mu} \, \mu > 0, 0 < t < \infty$ so that $R = \exp(-t_0 / \mu)$.

In the first procedure it is possible to observe the number surviving up to some test time $t_0' < t_0$ and to derive an estimate of R that is asymptotically unbiased. This procedure requires a single count of survivors at the predetermined time t_0'. Per contra, the second procedure requires a continuous monitoring of failure times and continues till a random time $t_{(n)}$, viz. the largest of observed failure times. In the first procedure, a random number of units is used up, while in the second this number is fixed as n. In the case of equipment which requires some form of energy for operation, the total operation time is also to be considered

With a complete sample of failure times, the sample mean of observed failure times is the minimum variance unbiased estimate of μ but the estimate $\widehat{R} = e^{(\frac{to}{t})}$ is asymptotically unbiased. So also is the estimate from the first procedure, viz. $\widehat{R1} = (\widehat{R})^{(t0'/t0)}$.

It comes out that $Var(\widehat{R}) = \dfrac{R(1-R)}{n}$ and $Var(R1) = Var(R)(\dfrac{to'}{\mu})^2$ so that the asymptotic relative efficiency of $\widehat{R2}$ relative to $\widehat{R1}$ is $E_1 = x^2 / (e^x - 1)$ where $x = t_0' / \mu$. Thus E_1 is independent of the mission time or the sample size and depends only on the test time. The maximum efficiency of the attribute plan compared to the variable plan is reached at the impractically high level of $t_0' = 1.6 \mu$.

We can alternatively choose $t_0' = E[t_{(n)} \mid \mu] = \Sigma\, 1 / (n - i + 1)$ to make the test times in the two procedures somewhat comparable. And note that the relative efficiency is now $E_1' = x_{(n)}^2 / [\exp\{x_{(n)} - 1\}]$ which diminishes monotonically with n, being highest, viz. 0.65, at n = 2. In this procedure $t_0' > 11\,\mu / 6$ for n > 3 and the attribute sampling plan is consistently less efficient than the variable sampling plan with the same (expected) test time, the relative efficiency being as small as 0.48 for n = 10. In a second bid to equate the two test times, one may compare the estimate based on failure times recorded till $t_0' = x$ along with the number still surviving with the estimate yielded by the attribute plan. We may either consider a sample of n failure times from an exponential distribution truncated at t_0' or use a censored sample of size n from the usual exponential. Deemer and Votaw (1955) provide estimates of their asymptotic variances. The asymptotic efficiency in the first case comes out as $E_2 = x^2 / [e + e^{-x} - 2 - x^2]$ while that for the second case can be shown to be $E_3 = x^2 / [e^x + e^{-x} - 2]$.

Evidently, $E_3 > E_2 > E_1$. Another possibility to be explored is to take the average of the random number of failure times by time t_0' (as prefixed for the attribute plan) and the average variance of estimated reliability over the density of this random number of failures can be compared with Var(R).

In view of the fact the total (expected) operating time in the attribute plan given by $T_1 = \Sigma\, E\,[t_{(i)}] + (n - r')\,t_0'$ where r' is the number of failures by the test time t'_0 and $t_{(i)}$ is the ith smallest failure time observed in the sample of n items differs from that in the variables plan, viz. $T_2 = \Sigma\, t_i = \Sigma\, t_{(i)}$, with the result that $T_2 > T_1$ for all r', it may be worth while to fix n_1 in the variables plan to have the same expected total operating time. On simplification of a somewhat involved algebra, it can be shown that $E\,(T_1) = n\,p$. Thus $n_1 = n\,p$ and this makes the efficiency in this case $E'' = E_1 / p = E_3$. It can be seen that for the usually employed low values of $x = t_0' / \mu$, efficiency is almost unity.

The average relative efficiency of the attribute plan with the same number of items used up comes out as $E''' = E_3$ again.

All this is just to stress the need to examine the implications of alternative plans for estimation of some parameter(s) of interest or testing some related hypotheses on effective sample sizes and other cost-influencing factors. Similarly, when using alternative prior distributions for parameters in some model, the different priors should be made comparable in respect of their means and variances by appropriate choices of hyper-parameters.

8.14 Multiple Comparison Procedures

Somewhat different from the comparison problems discussed earlier are problems in which multiple comparisons have to be made, not necessarily among different procedures or protocols or factor-level combinations but more often between two such entities at each of several end points or toll gates to facilitate an overall comparison between the two. Of course, a comparison among all to identify the best is of interest in many situations. And in this context we can proceed in terms of pair-wise comparisons. Multiple comparison problems have been discussed vigorously since the middle of the twentieth century. The related issue of multiple hypothesis testing also attracted the attention of researchers in the field of statistical inference.

In most of the multiple comparison procedures, our objective is to control the overall significance level for a set of inferences (usually as a follow-up of ANOVA). This overall significance level or error rate is the probability, conditional on all the null hypotheses under consideration being true, of rejecting at least one of them or, equivalently, of having at least one confidence interval not including the true value of the underlying parameter.

The alternative procedures available differ in the way these control the overall significance level and in their relative power. Duncan's 'multiple range test' does not control the overall significance level and is not generally considered as an alternative to the remaining procedures mentioned below.

Bonferroni: extremely general and simple, but often not so powerful

Tukey: best for all possible pair-wise comparisons when sample sizes are unequal or when confidence intervals are needed: very good even with equal sample sizes without confidence intervals being obtained

Step-down: most powerful for all possible pair-wise comparisons when sample sizes are equal

Dunnett: quite good to compare the 'control' group with each of the others, but not suitable for comparisons among the others

Hsu: compares each mean to the 'best' (largest or smallest as appropriate) of the other means

Scheffe: effective for unplanned contrasts among sub-sets of means

Sometimes, more than one approach can be followed and the one that proves to be the most efficient in the problem context and for the given data may be judiciously accepted. Benjamini and Hochberg (1995) pointed out some difficulties with these classic multiple comparison procedures that have led to their under-utilization. Procedures that control the family-wise error rate (FWER) at levels used in conventional single comparison problems tend to be much less powerful compared to the per-comparison procedures at the same level. A per-comparison error rate approach is probably recommended in such situations. Often the control of FWER is not really needed. From this angle, the expected proportion of errors among the rejected hypotheses – termed the false discovery rate – is a better yardstick. Since rejection of a hypothesis implies to an extent finding something different, the term 'false discovery rate' (FDR) used by Soric (1989) is justified. Benjamini and Hochberg discuss some Bonferroni-type FDR controlling procedure.

Table 8.7 may help in understanding FDR in a simple way, when we are interested in testing a known number m of (null) hypotheses, in terms of the numbers of errors we may commit.

It may be noted that R is an observable random variable (depending on sample observations and test procedures) while S, T, U and V are unobservable random variables. It may be noted that PCER = E (V / m) and FWER = Pr $(V \geq 1)$. If each hypothesis is tested at level α, then PCER $\leq \alpha$ and if the former is taken as α / m then FWER $\leq \alpha$. FDR is defined

TABLE 8.7

Number of Errors Committed in Different Situations

	Declared Non-significant	Declared Significant	Total
Null hypotheses true	U	V	m_0
Null hypotheses not true	T	S	$m - m_0$
Total	$m - R$	R	m

as $Q_e = E(Q) = E(V / R)$. If a procedure controls FDR only, it can be less stringent and a gain in power may be expected. Any procedure that controls the FWER also controls FDR. The following procedure to control FDR was proposed by Hochberg and Benjamini and later modified by them on the basis of some related findings by Simes (1986) and Hommel (1988).

Let $p_1, p_2, \ldots p_m$ be the p-values associated with a procedure to test the hypotheses $H_1, H_2, \ldots H_m$ respectively. Let $p_{(1)}, p_{(2)}, \ldots p_{(m)}$ be the ordered p-values and let $H_{(i)}$ be the hypothesis corresponding to $p_{(i)}$. Consider the following test procedure

Let k be the largest I for which $p_{(i)} \leq i / m \, q^*$ and reject all $H_{(i)} I = 1, 2, \ldots k$.

For independent test statistics for the different hypotheses and for any configuration of false null hypotheses, it can be shown that this procedure controls FDR at q^*. Simes had suggested a procedure to reject the intersection hypotheses that at least one of the null hypotheses is false if, for some I, $p_{(i)} \leq i \, \alpha / m$. Hochberg subsequently modified his procedure as:

Let k be the largest i for which $p_{(i)} \leq i \, \alpha / (m - i + 1)$, then reject all $H_{(i)} i = 1, 2, \ldots k$.

There have been several modifications of these test procedures and interesting useful applications of such multiple hypothesis testing problems have been reported in the literature.

The growing interest in multiple hypothesis testing problems owes a lot to the challenges posed by data analysis in new biology fields, such as proteomics, functional magnetic resonance imaging, brain mapping or genetics. And it is well known that data in these applications are strongly correlated (spatially or over time or over both time and space). It has been generally found that a positive dependency of certain types leads to a conservative behaviour of classic multiple test procedures. In fact, Sarkar (1998) showed that multivariate total positivity of order 2 (MTP_2) is particularly useful for control of FWER. This result allows an improvement over the closed Bonferroni test.

If strict control of FWER is targeted even with 'effective number of tests', we have low power for detecting true effects. This led Benjamini and Hochberg to relax the type I error criterion and to control FDR. Positive dependency in the sense of MTP_2 ensures FDR control of the linear set-up test proposed by Benjamini and Hochberg (Dickhaus and Stange, 2013). Dickhaus and Stange mention that the question of incorporating the concept of 'effective number of tests' in the LSU test is an open problem. They also point out that the question of how to modify the 'effective number of tests' proposed by them in cases where the correlation structure has to be estimated from the data remains a research problem.

8.15 Impact of Emotional Intelligence on Organizational Performance

One of the important factors influencing the performance of an organization, as recognized in all the business excellence models, is leadership, and leadership is comprehended in this context as a bunch of attributes possessed and displayed by leaders. An important attribute of leadership is emotional intelligence. To examine the role of emotional intelligence of business or operational heads or, generally speaking, decision-makers regarding process management, resource management and people management, we need some emotional intelligence inventory that can be applied to the decision-makers. We also need some acceptable measure of organizational performance. Among possible different measures of performance, financial results are quite important and we have to work in terms of some measures on which comparable data pertaining to different organizations can be conveniently obtained.

The above dependence analysis has been extended further to include some measures of labour and of capital inputs along with the EI factors as independent explanatory variables. In fact we have attempted a generalization of the traditional Cobb-Douglas production function equation to incorporate the concept of human capital (as one aspect of intellectual capital) in terms of emotional intelligence.

Several variations of dependent relations have been tried out given that labour and capital supplies can provide quite a reasonable explanation of variations in the dependent variables (either value addition or sales output). It is but natural that the additional explanation (percentage of variation in the dependent variable explained) by including EI factors will be somewhat low. We should not expect a significant increase in the coefficient of determination if we incorporate the impact of emotional intelligence. However, it is worthwhile to examine how much more we can explain variations in any dependent variable and we can also examine how different procedures to combine the EI score with 'labour' input can yield different results.

Examining the linear correlation of value addition as well as of sales turnover taken as dependent (output) variables with two traditional input (explanatory variables), namely net block and employee cost, along with the four components of emotional intelligence, introduced in this study, it is found that both for sales turnover and for value addition inclusion of EI components increases the multiple correlation.

In fact, the coefficient of determination (R^2) for sales turnover in the case of net block and employee cost as the dependent variables comes out as 0.74, while the same increases to 0.79 when the four EI components are added. Similarly, in the case of value addition, this coefficient increases from 0.85 to 0.88.

To make use of the emotional intelligence of decision-makers as an input component determining organizational performance, we thought of several ways of modifying the Cobb-Douglas production function.

In fact we wanted to change the classic labour input by a modified intellectual input with a human capital which takes care of the emotional intelligence attributes of leaders/decision-makers. Here again, we can use as a multiplying factor by which to adjust the employee cost figure

- either the average score of the four EI factor scores (averaged over the individuals within an organization) divided by 7 (the maximum possible score)
- or the consecutive product of the four EI factor scores divided by 7^4, equal to 2401.

Thus, the second independent input variable in the modified production function equation can be taken as

$$\frac{\text{Average EI scores}}{7} \times \text{Employee Cost} \quad \text{or} \quad \frac{\text{Product of EI scores}}{7^4 \quad \text{or} \quad 2401} \times \text{Employee Cost}$$

Taking sales turnover as the dependent output variable R comes to 0.76 with the first choice and 0.74 with the second choice. Similarly, considering value addition as the dependent output variable, R comes to 0.87 with the first choice and 0.81 with the second choice.

TABLE 8.8

Linear Multiple Correlation Values

Sl No.	Dependent Variable	Independent Variable(s)						R^2
1	Sales Turnover (S T)	Net Block (N B)	Employee Cost (E C)	-	-	-	-	0.74
2	Sales Turnover	Net Block	Employee Cost	Well-Being (W B)	Self-Control (S C)	Emotionality (Emo)	Sociability (Soc)	0.79
3	Sales Turnover	Net Block	Avg. of EI Factors / 7 × E C	-	-	-	-	0.76
4	Sales Turnover	Net Block	Multiplication of EI Factors / 2401	-	-	-	-	0.74
5	Value Addition	Net Block	Employee Cost	-	-	-	-	0.85
6	Value Addition	Net Block	Employee Cost	Well-Being	Self-Control	Emotionality	Sociability	0.88
7	Value Addition	Net Block	Avg. of EI Factors / 7 × E C	-	-	-	-	0.87
8	Value Addition	Net Block	Multiplication of EI Factors / 2401	-	-	-	-	0.81

TABLE 8.9

Weighted Correlations

Sl No.	Dependent Variable	Independent Variable						R^2
1	Sales Turnover	Net Block	Employee Cost	Well-Being	Self-Control	Emotionality	Sociability	0.85
2	Value Addition	Net Block	Employee Cost	Well-Being	Self-Control	Emotionality	Sociability	0.93

It should be noted that the traditional explanatory variables are already established as providing a reasonably high explanation of variations in output. The increase in R^2 with the introduction of EI components may appear to be small, but should not be regarded as trivial.

As shown in the Table 8.8, when the number of employees within an organization was taken as a weight for the variable values, the value of R^2 for sales turnover with all the six explanatory variables came out as 0.85, as against 0.79 when these weights were not taken into account.

Similarly the R^2 value increased from 0.88 to 0.93 in the case of value addition when weights were introduced into the computations.

It must be pointed out here that in this exercise, we have simply considered the emotional intelligence of decision-makers to modify the traditional 'labour' component in a production function approach to link output with labour and capital inputs. However, 'intellectual capital' or 'EI-adjusted labour force' should possibly embrace all workers, not remaining restricted to leaders or decision-makers only. Hence results obtained in the study reported here are just indicative of what can be done in such a context.

Principal-Component Analysis (PCA)

As observed in the table above, emotional intelligence comprehended in terms of the four components, viz. well-being, self-control, emotionality and sociability, provides a good explanation of variations in both sales turnover and value addition. To find out the relative contributions of these four components, one can work out a principal-component analysis and examine the loadings (coefficients) of each component in the first principal component (which is a linear combination of the four component scores such that the sum of variances of these four in the data set is mostly explained by this combination). We can possibly examine coefficients in the second principal component also.

9

Multivariate Analysis

9.1 Introduction

As has been pointed out in earlier chapters, most – if not all – sample surveys are multivariate and though designed experiments generally take into account a single response variable from each experimental unit, multi-response experiments are quite common in industry. In fact, it may be difficult to think of the objective in a research programme that can be achieved by analysing data on a single variable or several unrelated variables. It may be incidentally added that simply multiple data with little relation connecting the variables do not qualify to be branded as 'multivariate'.

In many cases, units relate to geographical locations or regions or to industrial units or clusters of such industries. The characteristics could relate to concentrations of several pollutants in air or water, prices of supplementary or complementary items of consumption, GDP along with FDI, exchange rate, rate of inflation etc. and so on. We may consider these figures for several countries in the same year or for one given country during several consecutive years.

Further, the different characteristics may really be the same characteristic noted for a unit on different occasions (points or periods of time) or in different locations or contexts. For example, we may have figures on the concentration of suspended particulate matter noted in different seasons at different locations, residential and commercial.

It should be noted that the different characteristics noted in respect of different units (howsoever defined) are not unrelated one to the others. In fact, these variable characteristics are expected to vary jointly or in sympathy, maybe owing to their dependence on some common underlying features of the units.

In fact, a common starting point to present multivariate data is the variance-covariance matrix or the inter-correlation matrix. With p characteristics recorded for n units, we have a data matrix of order n × p and a covariance matrix of order p × p. A second representation could be a set of n points in a p-dimensional space. With a suitable measure of distance or dissimilarity between any two units (based on the p characteristics) we have n(n − 1) / 2 such measures as the starting point.

Usually, the number of units and also the number of characteristics noted for each such unit are pretty large. The different characteristics may be expressed in different units and cannot be directly combined. These characteristics are inter-related among themselves, to different extents and – maybe – in different directions. These could definitely have different bearings on the underlying phenomenon/process.

If care has been taken to collect data so that they qualify as evidence throwing light on the phenomena being studied, it would be desirable to take account of each and every characteristic noted. However, to integrate collective features of all these characteristics and to derive some conclusions or even to get a reasonable description of the underlying phenomenon, large volumes of multivariate data would pose genuine problems.

Similarly, we can collect data in respect of a large number of units, e.g. when the experiment is non-destructive or non-invasive and not too costly to allow many repetitions. It is quite possible that these units may reveal similarities among units within some sub-groups – to be identified – and different such internally homogeneous sub-groups may justify different treatment subsequently. In such cases, there could be a need to reduce the initially large aggregate of units to a handful of homogeneous sub-groups of units.

These two previous aspects of multivariate data analysis, viz. reduction in dimensionality of the data or, put simply, in the number of variables to be analysed and dis-aggregating the initial group of units into sub-groups, are very important.

In terms of analysis, a dominant part is to investigate inter-relations among the characteristics, through relevant correlation and regression analysis. Multivariate linear or non-linear regressions are often used for prediction purposes beyond explaining a large part of the observed variation in the dependent variable(s). And dimension reduction is to be a preceding step for this task in some cases.

Beyond this is the need to comprehend the total variation (variance) in the data matrix in terms of an appropriate number of artificial variables derived as uncorrelated or independent combinations – usually linear – of the original variables, followed by attempts to associate meanings with these combinations. Two or three such combinations taken in order of their contributions to the total variance (of the original variables) – referred to as the first, second or third principal component – will serve the purpose. In fact, in the case of a large number of independent or explanatory variables in a regression analysis, we can work in terms of such linear combinations and use the method of partial least squares, if needed in the case of sparse data.

The other way round, we may need to understand each of the variables noted as a composite of some (relatively few) basic explanatory factors which have distinct connotations and are otherwise clearly understood. We may need to involve all these three aspects of analysis in a real-life problem. It should be noted that initially we carry out exploratory analysis without speaking about any model. Subsequently, we may proceed with confirmatory data analysis, assuming some model to represent the joint random variations among the characteristics contained in the given data set.

While principal-component analysis is primarily exploratory and model-free, factor analysis is model-based and confirmatory. Similarly, we have discriminant analysis – usually involving linear discriminant functions – to assign an unknown unit to one of several groups, which are first distinguished among themselves based on 'training' data sets. This also is model-based and confirmatory in nature.

We also have tools for clustering of data into 'homogeneous' groups or clusters, which can be regarded as 'units' in some subsequent analysis. Clustering is the task of assigning a set of objects or individuals into groups (called clusters) so that the objects in the same cluster are more similar with respect to some characteristic and in some sense to each other than to those in other clusters. Clustering is an important element in explorative data mining and a common technique for statistical data analysis in many fields, including machine learning, pattern recognition, image processing, information retrieval and bio-informatics.

Cluster analysis can be performed by various algorithms that differ significantly in their notion of what constitutes a cluster and how to find them efficiently. Popular options of clusters include groups with low distances between the cluster members, dense areas in the data space, intervals or particular statistical distributions. The clustering algorithm and the selection of parameters like the distance function to be used, a density threshold or the number of expected clusters depend on the particular data set and the intended use of results. Clustering is essentially an iterative process and a wide variety of algorithms is available Thus we have the aggregative or divisive methods of clustering. In the first, we start with any one object (usually one with a likely set of values for the multiple characteristics) and try to find out which of the remaining has the lowest distance from the first one. This one is put in the first cluster along with the starting object. A second cluster is now formed by considering the object in the remaining set which has the lowest average distance from the two objects in the earlier step l.

In this chapter, we discuss some important techniques for analysis of multivariate data, of course, to suit different objectives. The literature on each of these techniques has grown into full-fledged text and reference books. Though multivariate (multiple) regression analysis should have been included here, we omit this topic since some discussion has been attempted in the chapter on the role of modelling.

9.2 MANOVA

A direct extension of univariate ANOVA, MANOVA is a useful and, in some cases, essential tool in multivariate analysis – both in data-analytic and also in inferential procedures. Like ANOVA, this also requires the assumption of multivariate normality and homoscedasticity (common variance-covariance matrix). The second assumption can be checked by using Box's M-test conveniently.

For the sake of simplicity, we fist consider one-way classified data. The data on p variables for each of n individuals constitute a random sample, taken from each of k populations (corresponding to, say, k treatments). The data can be represented by the model

$$X_{ij} = \mu + \alpha_I + \varepsilon_{ij} \text{ where}$$

μ = overall mean vector (for the k populations)
α_I = vector of effects of the ith treatment and
ε_{ij} = vector of error components.
We assume, for the time being, effects α to be fixed and ε_{ij} to be distributed as $N(0, \Sigma)$.

The objective of analysis is to test the null hypothesis $H_0 (\alpha_1 = \alpha_2 + = \alpha_k)$ which is equivalent to the null hypothesis $H_0 (\mu_1 = \mu_2 = = \updownarrow_k)$ against the alternative H_A that at least one of the equations is not true. The likelihood ratio test statistic in this case is $\Lambda = |W| / |T|$, where $W = \Sigma W_i$ and $T_I = W + B$, W being the sum of squares within samples and B standing for the sum of squares between samples, given by $W = \Sigma W_I W_i = \Sigma_j (X_{ij} - X_{i \, bar})(X_{ij} - X_{i \, bar})'$ $I = 1, 2, \ldots k$ And $B = \Sigma_i n_i (X_{i \, bar} - X_{bar})(X_{i \, bar} - X_{bar})'$, $X_{i \, bar}$ being the mean of the ith variable based on n_i observations in the ith class and X_{bar} being the overall mean.

Under the null hypothesis, $W|/|T|$ has Wilk's distribution with degrees of freedom n, n – k, k – 1 and parameter $\prod_j (1 + \lambda_j)^{-1}$ where $\lambda_j = 1, 2, \ldots r$ are the eigenvalues of the matrix $W^{-1}B$. The null hypothesis is rejected for small values of the LRT test statistic. We can transform this statistic to F where

$$(n - k - p + 1)(1 - \sqrt{}) / p \sqrt{} = F \text{ with } p(k-1) \text{ and } (k-1)(n-k-p+1) \text{ d.f.}$$

9.3 Principal-Component Analysis

A prime concern in (empirical) research is to describe, measure, compare, analyse and explain (random) variations in some features characteristics of the perceptual world. The latter are revealed through observations on the outcomes of a repetitive experiment on some phenomenon/place in nature or the economy or society. Most often, we deal with several outcomes or random variables that are expressed in different units, are related among themselves to varying extents and have different bearings on the phenomenon under investigation.

To comprehend variations in a multivariate data set, we usually carry out an exploratory data analysis. Principal-component analysis (PCA) is one such analysis that attempts to reduce the dimensionality of the original data and to facilitate interpretation of the total variation in terms of a small number of synthetic variables, usually linear combinations of the original variables so derived that the loss of information (amount of unexplained variation) is the least possible. The original data set, viz. observed values of a p-dimensional random vector $X' = (X_1, X_2 \dots X_p)$, can then be replaced by the first two or three principal components for subsequent analysis.

Let the random vector $X' = [X_1, X_2, \dots, X_p]$ have the covariance matrix Σ with eigenvalues

$$\lambda_1 \geq \lambda_2 \geq \dots \lambda_p \geq 0.$$

Consider the linear combinations

$$
\begin{aligned}
Y_1 &= a_1'X = a_{11}X_1 + a_{12}X_2 + \dots + a_{1p}X_p \\
Y_2 &= a_2'X = a_{21}X_1 + a_{22}X_2 + \dots + a_{2p}X_p \\
&\dots\dots\dots\dots\dots\dots\dots\dots\dots\dots\dots\dots\dots \\
Y_p &= a_p'X = a_{p1}X_1 + a_{p2}X_2 + \dots + a_{pp}X_p
\end{aligned}
\tag{9.3.1}
$$

Then using (9.3.1), we obtain

$$
\begin{aligned}
\text{Var}(Y_i) &= a_i' \sum a_i & i &= 1, 2, \dots, p \\
\text{Cov}(Y_i, Y_k) &= a_i' \sum a_k & i, k &= 1, 2, \dots, p
\end{aligned}
\tag{9.3.2}
$$

The principal components are those *uncorrelated* linear combinations $Y_1, Y_2, \dots, Y_p$ whose variances in (9.3.2) are as large as possible.

The first principal component is the linear combination with maximum variance. That is, it maximizes Var $(Y_1) = a_i' \sum a_1$. It is clear that Var $(Y_1) = a_i' \sum a_1$ can be increased by multiplying any a_1 by some constant. To eliminate this indeterminacy, it is convenient to restrict attention to coefficient vectors of unit length. We therefore define

First principal component = linear combination $a_1'X$ that maximizes

$$\text{Var}(a_1'X), \quad \text{subject } a_1'a_1 = 1$$

Second principal component = linear combination $a_2'X$ that maximizes

$$\text{Var}(a_2'X), \quad \text{subject } a_2'a_2 = 1 \text{ and } \text{Cov}(a_1'X, (a_2'X) = 0$$

At the ith step,

Ith principal component = linear combination $a_i'X$ that maximizes Var $(a_i'X)$ subject to a_i' $a_i = 1$ and Cov $(a_i'X, a_k'X) = 0$ for $k < i$.

In this connection, the following results may be noted.

Result 1

Let Σ be the covariance matrix associated with the random vector $X' = [X_1, X_2,...,X_p]$. Let Σ have the eigenvalue-eigenvector pairs

$(\lambda_1, e_1), (\lambda_1, e_2), ... (\lambda_1, e_p)$ where $\lambda_1 \geq \lambda_2 \geq ... \lambda_p \geq 0$. Then the ith principal component is given by

$$Y_1 = e_1'X = e_{i1}X_1 + a_{i2}X_2 ... + a_{ip}X_p$$
$$i = 1, 2, ..., p$$

(9.3.3)

With these choices, $Var(Y_i) = e_i'\sum e_i = \text{»}_i x \quad i = 1, 2, ..., p$

$$Cov(Y_i, Y_k \neg) = e_k'\sum e_k = 0 \quad i \neq k$$

(9.3.4)

If some λ_i are equal, the choices of the corresponding coefficient vectors, e_i, and hence Y_i, are not unique.

Result 2

Let $X' = [X_1, X_2, ..., X_p]$ have covariance matrix Σ, with eigenvalue-eigenvector pairs

$$(\lambda_1, e_1), (\lambda_2, e_2), ... (\lambda_p, e_p) \text{ where } \lambda_1 \geq \lambda_2 \geq ... \lambda_p \geq 0.$$

Let $Y_1 = e_1'X, Y_2 = e_2'X..., e_p'$ be the principal components. Then

$$\Sigma_{11} + \sigma_{22} + ... + \sigma_{pp} = \sum_{i=1}^{p} Var(X_i) = \lambda_1 + \lambda_2 + ... + \lambda_p$$

$$= \sum_{i=1}^{p} Var(Y_i)$$

Consequently, the proportion of total variance due to (explained by) the kth principal component is

$$\lambda_k / (\lambda_1 + \lambda_2 + ... + \lambda_p) k = 1, 2, ... p$$

(9.3.5)

If most (for instance, 80 to 90%) of the total population variance, for large p, can be attributed to the first one, two or three components, then these components can 'replace' the original p variables without much loss of information.

Result 3

If $Y_1 = e_1'X, Y_2 = e_2'X..., e_p'X$ are the principal components obtained from the covariance matrix Σ, then

$$\rho Y_i, X_k = \frac{e_{ik}\sqrt{\lambda_i}}{\sqrt{\sigma_{kk}}} \quad i, k = 1, 2, ..., p$$

(8)

are the correlation coefficients between the components Y_i and the variables X_k. Here $(\lambda_1, e_1), (\lambda_2, e_2) \dots (\lambda_p, e_p)$ are the eigenvalue-eigenvector pairs for Σ.

Example 1 (calculating the population principal components). Suppose the random variables X_1, X_2 and X_3 have the covariance matrix

$$\Sigma = \begin{matrix} 1 & -2 & 0 \\ -2 & 5 & 0 \\ 0 & 0 & 2 \end{matrix}$$

It may be verified that the eigenvalue-eigenvector pairs are

$$\lambda_1 = 5.83, \quad e_1' = [.383, -.924, 0]$$
$$\lambda_2 = 2.00, \quad e_2' = [0, 0, 1]$$
$$\lambda_3 = 0.17, \quad e_3' = [.924, .383, 0]$$

Therefore, the principal components become

$$Y_1 = e_1'X = .383X_1 - .924X_2$$
$$Y_2 = e_2'X = .X_3 - .924X_2$$
$$Y_3 = e_3'X = .924X_1 - .383X_2$$

The variable X_3 is one of the principal components, because it is uncorrelated with the other two variables.

It is also readily apparent that

$$\sigma_{11} + \sigma_{22} + \sigma_{33} = 1 + 5 + 2 = \lambda_1 + \lambda_2 + \lambda_3 = 5.83 + 2.00 + 0.17$$

The proportion of total variance accounted for by the first principal component is

$$\lambda_1 / (\lambda_1 + \lambda_{2=} + \lambda_3) = 5.83 / 8 = 0.73.$$

Further, the first two components account for a proportion $(5.83 + 2) / 8 = 0.98$ of the population variance. In this case, the components Y_1 and Y_2 could replace the original three variables with little loss of information.

The variable X_2, with coefficient -0.924, receives the greatest weight in the component Y_1. It also has the largest correlation (in absolute value) with Y_1. The correlation of X_1 with Y_1, 0.925, is almost as large as that for X_2, indicating that the variables are about equally important to the first principal component. The relative sizes of the coefficients of X_1 and X_2 suggest, however, that X_2 contributes more to the determination of Y_1 than does X_1. Since, in this case, both coefficients are reasonably large and they have opposite signs, we would argue that both variables aid in the interpretation of Y_1.

9.4 Factor Analysis

Factor analysis tries to explain joint variations, represented by inter-correlations among observed, correlated variables in terms of a potentially fewer number of unobserved, uncorrelated variables, called factors. The observed variables are modelled as linear combinations of the potential factors, plus 'error' terms.

Computationally, this technique is equivalent to a low rank approximation to the matrix of observed variables. Factor analysis is related to principal-component analysis, but the two are not identical. Latent variable models including factor analysis use regression models to test hypotheses producing error terms, while PCA is generally an exploratory data analysis tool.

We have a vector X of p random variables observed for each of n individuals, with the mean vector $\mu = (\mu_1, \mu_2, ..., \mu_p)'$ and a covariance matrix $\Sigma_{p \times p}$.

The model used in factor analysis assumes linear dependence of these variable on a fewer number k of unobservable random variables, called common factors, and p additional sources of variation, called errors or specific factors. This linearity assumption is essential to maintain conformity with the covariance structure given by the model, which is stated as X (px1) – μ (px1) = L (pxk) F (kx1) + ϵ (px1) where X (pxn) is the set of n observations on each of p random variables, L (pxk) is the matrix of loadings of the p variables on the k unobserved factors, F (kxn) is the matrix of factor scores and ϵ (pxn) corresponds to errors. Each column of X and of F corresponds to one particular unit/individual/observation. Sometimes, the factors in F are common factors, while ϵ takes into account unique factors. In some analysis, μ is taken to be zero, and is interpreted as effects of a 'general factor' in the language of Thurstone. We assume that

1. F and ϵ are independent
2. E (F) = 0 and
3. Cov (F) = I (identity matrix)

Further, the components of ϵ are independently distributed with zero means and finite variances. Thus E (ϵ) = 0 and Cov (ϵ) = Ψ (matrix with only diagonal elements non-zero).

Expanding the model, we can consider the ith row X_i representing scores of the n individuals on observed variable i as a linear combination of unobservable factors and their loadings in the form $X_i = \Sigma l_{ij} f_j + e_i$ () where f_j is score on factor j, l_{ij} is the loading of the ith variable on the jth factor and e_i is the associated error factor. Estimators of factor j can be obtained as linear combinations of observed variables X_i in the form

$$f_j = w_1 X_1 + w_2 X_2 + ... + w_p X_p$$

The equation looks like a linear regression equation where each factor is expressed as a linear combination of q factors. However, the factors cannot be observed in a survey. It can be easily seen that $V(Xi) = \Sigma_j l_{ij}^2 + h_i$, the first part, is called communality.

If we now consider a data matrix $Z = C X$, where C is a diagonal matrix with elements C_i, the factor model can be written as Z = CX = C LF + C ϵ where V (Z) = C Σ C'.

Thus factor loadings of the scaled variables in Z are just C times those of the original variables X_i, showing the invariance of a factor analysis model.

Important issues in a factor analysis exercise include (1) finding an adequate number of factors, (2) estimation of factor-loadings, (3) estimation and interpretation of factors and (4) estimation of factor scores for the individuals.

Estimation of factor loadings eventually boils down to estimation of correlation coefficients between observed variables and unobservable factors.

Rotation of factors is done to ensure that (1) any column in the factor loading matrix should be close to zero, (2) entries in any row should have minimum values far from zero and (3) any two columns should reveal different patterns in respect of high and low loadings. And these can be attained if the original orthogonality of factors is retained through orthogonal rotation of factors. Important among orthogonal rotation of factors are (a) varimax rotation, (b) quartimax rotation and (3) equamax rotation. Oblique rotation in terms of algorithms for oblimax, quartimin, covarimin, is rarely used.

Through varimax rotation, a minimum number of variables get higher loadings. Further, communalities are not changed, nor the amount of variance explained by the components. This rotation does not influence the test of adequacy of the factor model. The factor correlation matrix becomes the identity matrix.

Quartimax rotation transforms the higher loadings of variable to loadings of moderate sizes and minimizes the number of factors needed to explain each variable. The resulting factor structure is not usually helpful to research workers. Equamax rotation is a mixture of varimax and quartimax rotations.

Quartimin, promax and oblimin rotations do not preserve orthogonality. and allow factors to be correlated, leading to diminished interpretability of the factors.

Beyond its initial uses in psychometry, factor analysis has been quite helpful in market research by identifying the salient attributes of features consumers use to evaluate products belonging to a certain category. With data collected from a sample of potential customers required to rate different product attributes or features, the factors extracted can be used to construct perceptual maps and other product positioning devices. Several useful applications of factor analysis have been reported in physical sciences also; e.g. in groundwater quality management, we may require to relate the spatial distributions of different chemical parameters to possible different sources like the existence of mines or large waste-fills. Factor analysis has found application in micro-array analysis. And such varied applications have motivated the development of appropriate software.

9.5 Cluster Analysis

9.5.1 Generalities

As noted earlier, cluster analysis or better clustering is a task – and not an algorithm or a set of algorithms – that precedes application of some inferential or even descriptive tool to analyse a data set on a large number n of individuals, each characterized by several, p say, inter-related variables. For individual k we have the variable values $X_k = (x_{1k}, x_{2k}, \dots x_{pk})$ $k = 1, 2, \dots n$. Depending on the measure of distance between two objects (individuals), the definition of a cluster, the basis for cluster formation and even the method involved therein, cluster analysis opens up a whole lot of diversity.

The distance between two individuals k and m can be taken as the Euclidean distance, viz. $[\Sigma_i (x_{ik} - x_{im})^2]^{\frac{1}{2}}$ or the Manhattan distance defined as $\Sigma \mid x_{ik} - x_{im} \mid$ or its generalization, viz. the Minkowski distance $\Sigma_I \mid x_{ik} - x_{im} \mid^r$ (r an integer) or we can take $\max_i \mid x_{ik} - x_{im} \mid$ or even the number $\# (x_{ik} - x_{im})$. It is true, however, that most algorithms make use of Euclidean distance, since the others are not amenable to convenient algebraic treatment.

A cluster means just a set or group of individuals. However, different algorithms describe clusters differently, based on different models. Inherent is the implication that clusters are as homogeneous as possible and different clusters differ as much as possible. Connectivity or linkage models of various types make use of distance measures, centroid models represent each cluster by its centroid or medioid, distribution models use statistic distribution, density models locate clusters as dense connected regions in the data space, etc. In practice the first two models are mostly used. And there again linkage could be quantified in terms of an average or the maximum or the minimum.

Finer distinctions among clustering algorithms are worth theoretical interest. In hard clustering, each individual finally belongs to one and only one cluster, while in soft clustering an individual may belong to several clusters with different probabilities. This situation is somewhat analogous to the idea of a fuzzy number taking values in an interval as indicated by a membership function. Clustering need not cover all the given individuals, leaving out some as outliers. Clusters could be non-overlapping or can overlap when a child in hierarchical clustering is included in the parent.

An important distinction should be made between supervised and unsupervised clustering, the first in terms of a pre-specified number of clusters, while in the second the number of clusters depends on the data and the algorithm used.

9.5.2 Hierarchical Clustering (Based on Linkage Model)

In hierarchical clustering the given data are not partitioned into a set of clusters in one step. Instead, a series of partitions takes place, which may run from a single cluster containing all the points (individuals) to n clusters each containing a single point. Clustering could be agglomerative in a series of fusions of the n points into clusters, or could be divisive by separating the n points successively into finer groupings. Agglomerative procedures are more often used. Hierarchical clustering may be represented by a two-dimensional diagram known as dendrogram, which illustrates the fusions of fissions/divisions made at each successive stage of analysis. In a dendrogram, the ordinate shows the distance at which the clusters merge, while the abscissa represents the points (individuals) such that the clusters do not mix.

Depending on how linkage is computed, we can have a variety of clustering procedures (algorithms). Popular choices are single-linkage clustering (based on the minimum distances between points), complete linkage clustering (based on the maximum distance) and the unweighted pair group method with arithmetic mean (also known as average linkage clustering). Also used sometimes is Ward's error sum of squares method.

The hierarchical method of clustering may also be applied for clustering variables, rather than individuals or units. Of course, for that purpose we will use correlations as measures of similarity or proximity and not some usual measure of distance. A high correlation between two variables indicates that these are similar to some extent. Traditional linkage measures will not apply. Instead we can take the correlation between the first principal components corresponding to two clusters of variables or the canonical correlations.

9.6 Discrimination and Classification

In many scientific enquiries we need to find the group or class to which a particular unit/ individual belongs, based on the observed values of several characters (random variables). In some cases, the groups have been already formed and can be differentiated in terms of these characters (as has been observed for some members belonging to each group), the problem being to assign a new individual to one of these classes/groups. There exist other situations where such groups/classes have not been formed to start with and we have to form some groups on the basis of contiguity or distance among the individuals as revealed by values of the characters observed. The latter concerns the problem of clustering, while the former is that of discrimination among entities (animals belonging to different orders/families/species, satellite images corresponding to different land forms or plant species, MRI pictures of the human brain under different psychological disorders, etc.) based on certain observable predictor variables and assigning some new entity to any of the identified classes, following some rules.

Let $\Pi_1, \Pi_2, \ldots \Pi_k$ denote k different populations and let $p_1, p_2, \ldots p_k$ be the respective prior probabilities. Random samples of size $n_1, n_2, \ldots n_k$ on p predictor variables $X_1, X_2, \ldots X_p$ are available to discriminate among the k populations and to develop a rule to assign a new entity with observed variable vector x to any of the k classes. If the prior probabilities are not known, these can be estimated as est $(p_i) = n_i / n$ where $n = \Sigma n_i$. Three different rules have been recognized, viz. (1) maximum likelihood discriminant rule, (2) Bayes discriminant rule and (3) Fisher's discriminant function rule. Of course the third reduces to the first if we assume multivariate normality and homoscedasticity of the data on predictor variables.

9.6.1 Bayes Discriminant Rule

This can be easily explained in the case of two populations and the procedure can be conveniently generalized to the case of any k populations. Let us assume the two distributions of the random vector X to be known for each population. Define R_1 as the set of values x for which we classify entities as belonging to Π_1 and R_2 as the remaining set of x values for which the entities are assigned to population Π_2. Obviously the sets R_1 and R_2 are exhaustive and mutually exclusive. The probability of wrongly classifying a new entity as belonging to Π_2 when it really belongs to Π_1 comes out to be

$$\Pr(x \, \varepsilon \, R_2 \mid R_1) = P(2 \mid 1) = f_{R2} \, f_1(x) \, dx$$

where $f_I(x)$ is the joint density of x in population I, I = 1 and 2. Similarly the other probability of an observed value x coming from Π_2 wrongly assigned to the other population can be computed. We can now compute the misclassification probability matrix (sometimes also called the 'confusion' matrix) in the following manner

$$\Pr(\text{correctly classified in } \Pi_1) = \Pr(x \, \varepsilon \, R_1 \mid \Pi_1) \, \pi$$
$$\Pr(\text{incorrectly classified in } \Pi_1 = \Pr(x \, \varepsilon \, R_1 \mid \Pi_2) \, \pi_2$$

We can draw up the matrix of misclassification costs as follows

True population	Classified as belonging to	
	Π_1	Π_2
Π_1	0	C (2 | 1)
Π_2	C (1 | 2)	0

Thus the expected cost of misclassification associated with any classification rule comes out as ECM = C (2 | 1) Pr (2 | 1) π_1 + C (1 | 2) Pr (1| 2) π_2.

The regions R_1 and R_2 that minimize ECM are given by the values of x such that R_1: $f_1(x) / f_2(x) > [C(1 \mid 2) \, \pi_2] / [C(1 \mid 2) \, \pi_1]$ and R_2 is the complementary region. If C (1 | 2) = C (2 | 1) and $\pi_1 = \pi_2$ this rule simplifies to $f(x \mid \Pi_1) > f(x \mid \Pi_2)$ and implies allocation of entity with x to population Π_1.

Under the assumption that the joint distributions in the two populations Π_1 and Π_2 are multivariate normal with mean vectors μ_1 and μ_2 respectively and a common variance-covariance matrix Σ, this rule simplifies to the rule: allocate individual x to the first population if

$$(\mu_1 - \mu_2)' \Sigma^{-1} x - \tfrac{1}{2} (\mu_1 - \mu_2)' \Sigma^{-1} (\mu_1 + \mu_2) \geq C (1 \mid 2) \pi_2 / C (2 \mid 1) \pi_1$$

Assign x to the second population otherwise. If we can assume $C (1 \mid 2) = C (2 \mid 1)$ and $\pi_1 = \pi_2$, when no other information is available about these relations, this rule can be operationalized as follows: allocate the entity with observed predictors x to Π_1 if

$$(m_1 - m_2)' S_u - !x - \tfrac{1}{2} (m_1 - m_2)' S_u^{-1} (m_1 + m_2) \geq 0$$

where m_1 and m_2 are the sample mean vectors for populations 1 and 2 respectively and S_u is the pooled sample variance-covariance matrix. In fact, the left-hand-side function is referred to as the linear discriminant function.

If multivariate normality assumption is not valid and a common variance-covariance matrix cannot be assumed, Smith (1947) suggested a quadratic classification rule based on a quadratic composite of observed predictors estimated in terms of

$$Q (x) = \tfrac{1}{2} \ln \mid S_2 \mid / \mid S_1 \mid - \tfrac{1}{2} (x - m_1)' S_1^{-1} (x - m_1) + \tfrac{1}{2} (x - m_2)' S_2^{-1} (x - m_2)$$

where S_1 and S_2 are the sample variance-covariance matrices for the two samples. The entity is assigned to Π_1 if $q (x) > 0$, otherwise it is assigned to Π_2.

9.6.2 Fisher's Discriminant Function Rule

Fisher defined the discriminant function between two populations as that linear function of the characters (variables) for which the ratio

$$(\text{mean difference})^2 \div \text{variance}$$

is a maximum. Let $l_1 X_1 + l_2 X_2 + \dots + l_p X_p$ be the linear function and ∂_i be the difference between the mean values of X_i in the two populations. Then the quantity to be maximized is

$$(\Sigma \, l \, i \, \partial \, i)^2 \div \Sigma\Sigma \, l \, i \, lj \, \sigma \, ij$$

With k populations, let B be the sum of products and squares matrix and W be the sum of within sums of products and squares. Then we take $D = b'X$ as the linear discriminant function. Fisher suggested that b be so determined that the ratio $\lambda = b'Bb / b'Wb$ is a maximum. This implies $(B - \lambda W) b = 0$. This means that λ is the largest eigenvalue of $W^{-1} B$ and b is the eigenvector corresponding to the largest λ.

The mean discriminant score for the ith population is $D_i = b'm_i$, where m_i is the sample mean vector for the ith population. Hence given the vector x of p characteristics of a new individual, we assign this individual to population (class) r if $\mid b'x - b'm_r \mid < \mid b'x - b'm_i \mid$ for all i not equal to r.

It is easy to see that Fisher's discriminant rule is equivalent to the ML discriminant rule in the case of two populations provided $\pi_1 \sim N_p (\mu_1, \Sigma)$ and $\pi_2 \sim N_p (\mu_2, \Sigma)$. For two groups with training samples of sizes n_1 and n_2 and discriminant scores D_1 and D_2, the point of separation becomes $D^* = (n_1 D_1 + n_2 D_2) / (n_1 + n_2)$.

9.6.3 Maximum Likelihood Discriminant Rule

If $L_i(x)$ is the likelihood of the sample observation x then the maximum likelihood discrimination rule would state x belongs to class h if $L_h(x) = \max L_i(x)$. For $p = 2$ and π_1 denoting $N(\mu_1, \sigma_1^2)$ and π_2 denoting $N(\mu_2, \sigma_2^2)$. For the case of equal variances, $L_1(x) > L_2(x)$ if $|x - \mu_1| > |x - \mu_2|$. Thus if $\mu_2 > \mu_1$, the unit with observation x will be assigned to π_2 if $x > |\mu_1 + \mu_2|$. The rule will be reversed if $\mu_2 < \mu_1$.

The above rule can be directly generalized to the case of several multivariate normal populations $\pi_i (\mu_i, \Sigma_i)$ i = 1, 2, ... k. In the case of identical dispersion matrices with a common dispersion matrix Σ the above rule would assign the observation x to the class (population) for which the Mahalanobis distance $(x - \mu_i) \Sigma^{-1} (x - \mu_i)$ is a minimum.

Let us illustrate this rule by a simple example with samples drawn from two trivariate populations with est $\mu_1 = (4.821, 0.500, 7.643)'$, est $\mu_2 = (3.625, 0.375, 9.312)$ and dispersion matrices

$$S_1 = \begin{vmatrix} 4.416 & 1.017 & -5.992 \\ 1.017 & 0.607 & -1.428 \\ -5.992 & -1.428 & 24.729 \end{vmatrix} \text{ and } S_2 = \begin{vmatrix} 4.359 & 1.453 & -5.008 \\ 1.453 & 0.984 & -1.367 \\ -5.008 & -1.367 & 18.339 \end{vmatrix}$$

The two covariance matrices can be tested to be homogeneous and so we have the pooled dispersion matrix (with the three rows written side by side)

$S_u = [4.425\ 1.232\ -5.902\ |\ 1.232\ 0.780\ -1.473\ |\ -5.902\ -1.473\ 23.472]$. We also compute b $- S_u^{-1}$ (est μ_1 −est μ_2) = [0.4036, −0.4761, 0.0004].

Let us consider an individual with the observation $x = (6, 1, 9)$. Using the above ML rule, this observation should be assigned to population 1.

9.6.4 Classification and Regression Trees

Classification and regression trees (CART) refer to prediction models based on data comprising a dependent (response) variable y and a set of p independent explanatory variables. $X_1, X_2, ... X_p$ through recursive partitioning of the covariate space resulting in tree structures. If the response variable is categorical with a finite number of unordered values 1, 2, ... k this algorithmic procedure results in a classification tree, whereas if y is a quantitative character taking numerical values we have regression trees. The basic idea is to predict the value of y for a new set of explanatory variables. using a general regression-like model as follows

$$y = \varphi(x_1, x_2, ... x_p) + e$$

The variable y could be binary with values 0 and 1 and we could consider its logit transform z = log [Prob (y = 1) / {1 − Prob (y = 1)}] while x-values could be both categorical and numerical. In the case of a binary y, we can use the above relation to assign a new unit (a new set of x-values) to either of two classes as in the discriminant problem considered earlier. The covariates or explanatory variables could be interacting among themselves and the relation in the equation stated above could be non-linear. Treating the sample data as the training data set, CART following some algorithms illustrates a useful application of machine learning.

The procedure uses binary splits of the covariate space on the basis of a chosen splitting variable, say X_j, to yield a left node $S_1 = \{x: X_j \leq x\}$ and a right node $S_2 = \{x: X_j > x\}$. The objective is to find a split that ensures highest possible homogeneity in respect of the response variable y within a node and highest possible heterogeneity between the two sets. Each of the nodes is again split into a left and a right node on the basis of another splitting variable. Continued, this splitting eventually results in rectangular regions. In a regression tree, responses of all cases or units in the leaf nodes of the tree are averaged by taking the mean or the median and we get a piecewise constant regression function. The leaf nodes are the final nodes of the tree when no further splitting serves any purpose going by a certain criterion. Splitting stops either when the number of cases in a leaf node falls below a specified minimum or when certain p-values in the splitting criteria are greater than a pre-determined value.

A considerable volume of literature has grown this topic, including Bayesian CART approaches. Mention should be made of the contributions by Breiman et al. (1984), Chaudhuri et al. (1994, 1995), Loh (2002) and Denison et al. (1998), among several others. In the context of regression trees, an algorithm GUIDE (generalized, unbiased inter-action, detection and estimation) and some of its modified versions are in common use. One disadvantage of CART in this context is that poor or wrong splits in upper nodes near to the root node or at the root node itself can lead to poor results in nodes near to the leaf nodes.

For the purpose of regression trees, splitting proceeds on the basis of R^2 values for the regression at any stage. In the case of classification trees, misclassification probabilities and associated costs provide the criterion for fit and signal for continuation or stoppage.

9.6.5 Support Vector Machines and Kernel Classifiers

A support vector machine (SVM) is a supervised discriminatory data-analytic procedure for binary pattern recognition from the training data set and for the associated problem of classifying a test data set into one of two classes, developing a hyperplane maximizing separation from each of the two groups in the training data. SVMs are based on the concept of decision boundaries. An SVM constructs hyperplanes in a multidimensional space that separate cases of different class labels. SVM supports both regression and classification tasks besides the task of detecting outliers in data, and can handle multiple continuous and categorical variables. To construct an optimal hyperplane, SVM employs an iterative training algorithm which is used to minimize an error function. Different error functions are chosen for regression and classification tasks. To facilitate computations, dot products of pairs of input data may be obtained easily in terms of variables in the original space and for this a kernel mapping K (x, y) is used. The hyperplanes are defined by vectors which are linear combinations with parameters α_i of images of feature vectors x_i that are given in the data base.

SVM has often been used to classify new objects with multiple features in either of two classes, like defective and non-defective articles, based on a number of measurable quality characteristics for each article, sick or healthy based on several health-related parameters, and so on. In addition to performing linear classification, SVMs can effi-ciently perform non-linear classification by using the kernel trick, implicitly mapping their inputs into high-dimensional feature spaces. Some commonly used kernels to deal with non-linear classification problems are of the following types; polynomial (homo-geneous), polynomial (inhomogeneous), hyperbolic tangent and Gaussian radial basic function.

9.7 Multi-Dimensional Scaling

9.7.1 Definition

Multi-dimensional scaling (MDS) is a method to display (visualize) the relative positions of several objects (subjects) in a two- or three-dimensional Euclidean space, to explore similarities (dissimilarities) revealed in data pertaining to the objects/subjects. Such similarities (dissimilarities) refer to pairs of entities, as judged objectively in terms of physical parameters or assessed subjectively in terms of opinions. An MDS algorithm starts with a matrix of item-item similarities and ends in assigning a location to each item in a multi-dimensional space. With two or three dimensions, the resulting locations may be displayed in a graph or 3D visualization. Also known as Torgerson scaling, MDS is a set of statistical techniques used in information visualization

In some sense, multidimensional scaling, introduced by Torgerson, may be regarded as an extension of product scaling, introduced by another psychologist, Thurstone. In product scaling, we consider a number k of concrete entities which are presented pair-wise to a panel of n judges, each of whom is required to prefer one entity within a pair to the other in terms of some pre-specified features. The final data appear as a $k \times k$ matrix in which the (i, j) th element is p_{ij} = proportion of judges preferring entity j to entity i. Using the Law of Comparative Judgement and the method of discriminal dispersion, a scale value is finally found for each entity so that their relative positions can be shown as points on a real line. In multi-dimensional, the entities are displayed as points in a two- (or at most three-) dimensional Euclidean plane

9.7.2 Concept of Distance

If we start with similarities or affinities δ_{ij} between two objects or entities i and j, we can deduce dissimilarities d_{ij} by choosing a maximum similarity $c \geq \max \delta_{ij}$ and taking $d_{ij} = c - \delta_{ij}$ for i and j different and is zero otherwise. One apparent problem will arise with the choice of c, since the ultimate picture will vary from one choice to another, which is an undesirable situation. However, non-metric MDS takes care of this problem and even CMDS and Sammon mapping fare pretty well in this context.

Distance, dissimilarity or similarity (or proximity) is defined for any pair of entities (objects) in space. In mathematics, a distance function (which provides a distance between two given objects) is also called a metric d (X, Y), satisfying

$$d (X, Y) \geq 0$$
$$d (X, Y) = 0 \text{ if and only if } X \equiv Y$$
$$d (X, Y) = d (Y, X) \text{ and}$$
$$d (X, Z) = d (X, Y) + d (Y, Z)$$

A set of dissimilarities need not be distances or, even if so, may not admit of interpretation as Euclidean distances.

Similarities can be converted into dissimilarities by using the formula $d_{ij} = \sqrt{[s_{ii} + s_{jj} 2 s_{ij}]}$ where s_{ij} is a measure of similarity, while d_{ij} is a measure of dissimilarity.

The starting distances could be physical distances (driving or flying) between cities, taken in pairs, within a region like a continent or a big country, as are shown by points on the plane of a Cartesian map or on the surface of a globe. These distances are those on a non-Euclidean surface. They could be absolute differences in ratings or ranks.

Starting with a matrix of distances or dissimilarities $D = ((d_{ij}))$ among pairs of n objects or subjects (to be compared) in respect of some feature or features, MDS comes up with a matrix of estimated distances $D' = ((d'_{ij}))$, estimation being done to ensure the difference between D and D'' is as small as possible. MDS produces a map of the n objects usually in two dimensions so as to preserve the relative positions of the objects.

9.7.3 Classic MDS (CMDS)

In classic or metric MDS the original metric or distance is reproduced, whereas in non-metric MDS ranks of the distances are reproduced. Going by a two-dimensional map to be provided by MDS, each object will be eventually reproduced by a pair of coordinates or a point on a scatter plot from which we can visually assess the distance between the two points in each pair, which can be conveniently taken as the Euclidean distance $d_{ij}' = \sqrt{\sum (x_{ik} - x_{jk})^2}$ (summed over k = 1, 2) where x_{ik} and x_{jk} are the coordinates of the points corresponding to objects i and j.

To work out values of d_{ij} amounts to determining values of x_{ik} and x_{jk}, k = 1, 2, ... n or to find X so that $|| x_i - x_j || = d'_{ij}$ so that we get points on a two-dimensional plane in such a way that d'_{ij} is as close as possible to d_{ij}. The task is not just to find the distances d_{ij} but the coordinates. Such a solution is not unique, because if X is the solution, then $X^* = X + C$ satisfies $|| x_i^* - x_j^* || = || (x_i + C) - (x_j + C) || = || x_i - x_j || = d^*_{ij}$. Any location C can be used, but if we speak of a centred configuration with the constraint $\sum x_{ik} = 0$ over I, for all k. then we overcome this problem.

Since D is a positive definite matrix, we can find a Gram matrix $B = X' X$ (X assumed to be centred). This gives

$$d^2_{ij} = b_{ii} + b_{jj} - 2 b_{ij} \text{ from } || x_i - x_j ||^2 = x'_I x_i + x'_j x_j - 2 x'_i x_j$$

The centring constraint leads to

$$\sum_I b_{ij} = \sum_I \sum_k x_{ik} x_{jk} = \sum_k x_{jk} \sum_i x_{ik} = 0 \text{ for } j = 1, 2, n$$

We now have

$$\sum_I d_{ij}^2 = T + n b_{jj}, \sum_j d_{ij}^2 = T + n b_{ii} \text{ and } \sum_j \sum_I d_{ij}^2 = 2 n T$$

where $T = \sum b_{ii}$ is the trace of B. Now we have the unique solution

$b_{ij} = -1/2 (d_{ij}^2 - d_{i0}^2 - d_{0j}^2 + d_{00}^2)$ where d_{i0} and d_{0j} are the totals of row i and column j respectively, d_{00} being the grand total. A solution X is then given by the eigen-decomposition of B. Thus, for $B = V \Gamma V'$ we get $X = \Gamma^{1/2} V'$.

If we wish to visualize the distances on a two-dimensional plane, then the first two rows of X best preserve the distances d_{ij} among all other linear dimension reductions of X.

The above derivation in the case of a two-dimensional representation can be simply stated by way of the following algorithm:

1. From **D** calculate $\mathbf{A} = \{-1/2 \, d_{ij}^2\}$.
2. From **A** calculate $\mathbf{B} = \{a_{ij} - a_{i0} - a_{0j} + a_{00}\}$, where a_{i0} is the average of all a_{ij} across j.

3. Find the p largest eigenvalues $\lambda_1 > \lambda_2 > \lambda_3 \ldots > \lambda_p$ of B and corresponding eigenvectors $L = \{L_{(1)}, L_{(2)}, \ldots, L_{(p)}\}$ which are normalized so that $L_{(i)}/ L_{(i)} = \lambda_i$. We are assuming that p is selected so that the eigenvalues are all relatively large and positive.
4. The coordinate vectors of the objects are the rows of **L**.

9.7.4 An Illustration

Let us take the example considered by Mukherjee et al. (2018) in which a set of eight brands of TV sets were presented to a single judge, to get their performance ranks based on each of several aspects of performance such as picture clarity, sound control, ease of remote control and of channel change etc. We take into account the average rank assigned to any set. We can involve several judges and only the most important performance parameter for each TV set and take the average ranks. We can combine both the situations and try out replicated MDS.

It is also possible that for each set we get the proportion of judges who consider the given set as the best.

For the purpose of this illustration of classic MDS, suppose we have the average ranks as

Set	1	2	3	4	5	6	7	8
Rank	7	4	1	5	2	8	6	3

We can now construct the ABSOLUTE rank-difference or distance matrix as follows:

Set	1	2	3	4	5	6	7	8	
1	0	3	6	2	5	1	1	4
2	3	0	3	1	2	4	2	1
3	6	3	0	4	1	7	5	2
4	4	1	4	0	3	3	1	2
5	5	2	1	3	0	6	4	1
6	1	4	7	3	6	0	2	5
7	1	2	5	1	4	2	0	3
8	4	1	2	2	2	1	5	3	0

- We have to find the coordinates of eight points on a two-dimensional plane so that the matrix of Euclidean distances between pairs of points and the matrix just now obtained is a minimum. One simple way is to consider the sum of squared differences between these two distance measures over all possible pairs divided by the sum of the original distances as the criterion of fit and the best-fitting representation is the one in which this sum (stress) is minimized.

- Following the procedure indicated in section 9.7.3, we find the largest and the second largest eigenvalues as 41.75 and 5.64 respectively. The eight brands can now be represented by points on a Euclidean plane with coordinates (2.53, 1.54), (−0.46, 0.03), (−3.46, 0.34), (0.19, −1.71), (−2.46, 0.23), (3.55, −0.38), (1.55, −0.18) and (−1.45, 0.13). As can be easily appreciated, the points are quite well separated one from the other.

9.7.5 Goodness of Fit

Stress or normalized stress is the most widely accepted measure of goodness of fit and is defined as stress = $[\Sigma_{i<j} (d_{ij} - \delta_{ij})^2 / \Sigma_{i<j} d_{ij}^2]$ where d and δ correspond to the given (actual) and the reproduced distances respectively. Stress could also be defined by replacing the denominator by the sum of squared reproduced distances. The smaller the value of stress, the better is the fit.

It is obvious that MDS is able to reproduce the original relative positions in the map as stress approaches zero. Kruskal (1964) suggested the following advice about the stress values:

Stress	Goodness of fit
0.200	poor
0.100	fair
0.50	god
0.25	excellent
0.0	perfect

More recent articles caution against using such a table as this, since acceptable values of stress depend on the quality of the distance matrix and the number of objects and also the number of dimensions used.

9.7.6 Applications of MDS

Multi-dimensional scaling has since been applied in diverse fields, ranging from mundane problems of marketing to highly complex issues of affective and cognitive reasoning. In the problem-solving arena, if we have to face a problem we have never met before, we need to use our intuition. And intuition, in this context, can be explained as the process of making analogies between the current problem and the ones solved in the past to eventually find a suitable solution, which may not be the 'best'. Establishing such analogies or affinities or similarities must start with some features of a problem which may be metrics or otherwise and then to consider the closeness or difference in nature or magnitude of each feature between the current problem and the already solved one.

Osgood (1975) used multi-dimensional scaling to facilitate visualisation of affective words based on similarity ratings of the words presented to subjects representing different cultural backgrounds. For any pair of words, the similarity rating was obtained from a subject from one cultural background and the multiple distance matrices from the different cultural background representatives were used for MDS to get the averaged distances between words presented as points on a two-dimensional plane. The other possibility is to focus on inter-cultural differences in which case a distance matrix where each element could be the rank difference for a word between two cultural groups could be constructed and we can have as many distance matrices as the number of words.

MDS finds useful application in marketing. Potential customers are asked to compare pairs of products and to make judgements about their similarity. Whereas other techniques such as factor analysis, discriminant analysis and conjoint analysis obtain underlying dimensions of factors from responses to product attributes identified by the researcher, MDS obtains the underlying dimensions from respondents' judgements about the similarity or dissimilarity between products. It does not depend on researcher judgement. It does not require a list of attributes for a product to be presented to a potential customer.

Because of this, MDS is the most commonly used technique for perceptual mapping. For easy conceptualization of variation among objects or subjects in a group, we sometimes use multi-dimensional scaling to provide a visual two- (or three-) dimensional representation of the original data.

9.7.7 Further Developments

A major development over the classic MDS provides for several distance matrices to be incorporated with the provision that these could differ from one another in systematically non-linear or even non-monotonic ways. Weighted MDS (WMDS) is one step beyond replicated MDS to account for individual differences in the fundamental perceptual or cognitive processes that generate the responses. While RMDS generates a single distance matrix so determined that it is simultaneously like all the distance matrices, WMDS generates as many distance matrices as the starting ones, as close as possible to the starting ones.

Along with artificial neural networks and clustering techniques, multi-dimensional scaling is being used to comprehend human intelligence (sometimes branded as super-intelligence to solve difficult problems) in terms of distances perceived and quantified between the present state of a system and its past. Complex affective and cognitive processes are being analysed using some variants of multi-dimensional scaling.

10

Analysis of Dynamic Data

10.1 Introduction

Forecasting the future state of the underlying variable with a level of inaccuracy that can be determined (in terms of some measure of uncertainty like probability) is nearly mandatory in several human engagements – professional as well as scientific. In fact, the ability to forecast is crucial for planning and control. And phenomena to which the variable (scalar or vector) relates may be varying over time under various forces in a wide range of patterns from pretty regular and, say, linear through random to completely chaotic.

Dynamic data may arise in an altogether different context as well. Thus, in a physical experiment planned and conducted by some research worker, the response to a given treatment combination (a stimulus) may be time-dependent and may be observed over a time period till the response stabilizes. Thus, for each treatment combination, we get what may be taken as a time series on the response. For the sake of simplicity, we may consider a numerical response. The observed response paths may not exhibit the same patterns in terms of change points for the different treatment combinations. This situation is not the same as when we observe several such response paths, one for each response variable in a multi-response experiment, all for the same treatment combination.

Modern time-series data sets often defy traditional statistical assumptions. As pointed out by Fryzlewicz (2014), time-series data in may contexts are massive in size and high-dimensional in macro-economic models where many potential predictors are frequently included in models, e.g. for GDP growth or for mobility across occupations among members of a cohort non-normally distributed (e.g. in finance where daily returns on many financial instruments show deviation from normality) and non-stationary, which means that their statistical properties such as the mean, variance or autocovariance change with time (e.g. in finance where the co-dependence structure of markets is known not to remain constant throughout but to respond to financial crises). Time-series data also arise as complex patterns such as curves (e.g. yield curves). These and similar issues arising particularly in the context of multivariate time-series data call for new models and, beyond that, new techniques to offer satisfactory prediction.

Often, models are relations connecting the observed value of the variable with some unobserved or unobservable factors. Besides, some probability distribution(s) will be involved for the random variable. The idea is to reveal the nature and extent of the underlying time-dependence. Any model will involve several (unknown) parameters to be estimated from the given data, with or without some relevant prior information.

With a journal exclusively devoted to the topic and a vast literature growing around it, only a very few issues in time-series analysis, specially those which have not so far been widely applied but are useful in some research, have been briefly taken up in this chapter. In no way is the chapter meant to be a balanced overview of the multi-faceted subject.

10.2 Models in Time-Series Analysis

The selection of an 'appropriate' model is an important task in time-series analysis.

Popular 'structural' models are built in terms of components which admit direct interpretation, e.g. classic decomposition in terms of a trend, a seasonal and an irregular component, each being deterministic or stochastic. A stochastic model is a regression of the observed variable on explanatory variables like a trend and a set of seasonal dummies. Regression coefficients could be time-varying. Univariate structural models are usually regression models with explanatory variables as functions of time and parameters which vary with time. Adding observable explanatory variables is a natural extension, as is the construction of multivariate models.

As observed by Harvey and Shephard (1993), "the key to handling structural time series models is the state-space form with the state of the system representing the various unobserved components such as trends and seasonals. The estimate of the unobservable state can be updated by means of a filtering algorithm as new observations become available. Predictions are made by extrapolating these estimated components into the future, while smoothing algorithms give the best estimate of the state at any point within the sample".

All linear and time-invariant structural models can be reduced to ARIMA (auto-regressive integrated moving average) form. Thus a model given by $y_t = \mu_t + \varepsilon_t$ with $\mu_t = \mu_{t-1} + \delta_t$ where ε_t and δt are uncorrelated white noise disturbances can be reduced, on first differencing, as y_t corresponds to an ARIMA (1,1,1) model.

The starting point in conventional time-series modelling is a stationary stochastic process.

Non-stationarity is handled by transformation of variables like logarithms and differencing, leading to the ARIMA class. Stationarity can be tested.

If we need to incorporate observable explanatory variables, some prior knowledge about which variables should enter the model has to exist and be used.

A good guide to model selection is given by a plotted representation of the original series, or the series after a suitable transformation of data, or after suitable differencing in the original or differenced data, or of the auto-correlation function.

Some idea about the evolution of the time series and associated economic or other phenomena will always be useful.

10.2.1 Criteria for Model Selection

Parsimony judged in terms of fit to observed data and the number of parameters to be estimated – basically penalized likelihood (Akaike or the Bayesian information criterion)

Consistency with prior knowledge

Structural stability and a few others.

Let $L_n(k)$ be the maximum likelihood of a sample of n observations based on a model with k parameters. The model that minimizes $c_n(k)$, given below, is chosen.

$$c_n(k) = -2\ln(L_n(k)) + k\Phi(n) \text{ where}$$
$$\Phi(n) = 2 \qquad \text{in Akaike criterion}$$
$$= -2\ln(\ln(n)) \text{ in Hannan-Quinn criterion}$$
$$= \ln(n) \qquad \text{in Schwarz criterion}$$

In the ARMA (p, q) case, $c_n(k)$ reduces to

$$2n\ln\sigma + 2(1 + p + q) \qquad \text{Akaike}$$
$$2n\ln\sigma + 2(1 + p + q)\ln(\ln(n)) \text{ Hannan-Quinn}$$
$$2n\ln\sigma + (1 + p + q)\ln(\ln(n)) \text{ Schwarz}$$

where σ^2 is the ML estimator of the error variance. Numbers (p, q) are determined to minimize $c_n(k)$.

Similar expressions for the three criteria can be found for (G)ARCH (generalized autoregressive conditionally heterogeneous) models with k parameters.

The selected model has to provide a good fit to the observed data, say, in terms of some criterion like root mean square error or mean absolute error or coefficient of determination.

Fit could be examined over the entire period covered by the observed data. Otherwise, we could fit the model over a shorter period and use the 'hold-out' sample comprising the remaining data for the purpose of forecasting.

- Many economic and financial time-series undergo episodes in which the behaviour of the underlying variable changes dramatically.
- Financial panics, wars, major changes in policy like allowing FDI and/or raising the cap on it in some sectors of the economy, withdrawing subsidies, devaluing currency, changing interest rates, etc. affect many economic indicators.
- To fit a constant-parameter, linear time-series representation across such episodes invites a serious error.

Consider a Gaussian AR(1) process

$$y_t - \mu = \Phi(y_{t-1} - \mu) \, \varepsilon_t \text{ with } \varepsilon_t \sim \text{iid } N(0, \sigma) \text{ and } I\Phi I < 1.$$

Suppose at some point in the data, μ changed from μ_1 to μ_2.

To forecast, we need a (probability) law to model the change, e.g. if $y_{t-1} < 0$, take μ_1, take μ_2 otherwise (threshold models).

Changes may be due to processes largely unrelated to past realizations and not directly observable.

A Markov chain with two unobserved state variables S_1 and S_2 linked to μ_1 and μ_2:

$$\Pr(s_t = i \, / \, s_{t-1} = j) = p_{ij} \; i,j = 1, 2$$
$$\text{with } p_{11} + p_{12} = p_{21} + p_{22} = 1.$$

We now model the data as

$$(y_t - \mu_{s_t}) = \Phi(y_{t-1} - \mu_{s(t-1)}) + \varepsilon_t$$

A permanent shift in μ is admitted by $p_{21} = 0$ and its near-zero value an unlikely return to regime 1.

We allow the data to tell us the nature and incidence of significant changes.

We now have the probability density

$$p\ (y, y, \dots y; \lambda)\ \text{where}$$
$$\lambda = (\Phi,\ \sigma,\ \mu 1,\ \mu 2,\ p_{11},\ p_{22})$$

To find an estimate of λ, to form an inference about when changes occurred and, using these, to forecast the series and to test economic hypotheses are the tasks.

10.3 Signal Extraction, Benchmarking, Interpolation and Extrapolation

Whenever time-series data are involved in a research into some social or economic or even environmental phenomenon, the research worker may come across different series on the same variable released by different agencies with different periodicities, annual, quarterly, monthly, weekly etc. And different series may not be equally credible, some usually less frequently available ones may afford greater care and vigil in data collection, while others may not be in a position to exercise that much supervision to avoid missing values, biases and inaccuracies. In such cases, researchers often make adjustments to conveniently available data by way of (a) signal extraction, (b) benchmarking and (c) interpolation and extrapolation to increase the efficiency of estimates, reduce biases and estimate missing values. In fact, signal extraction or estimation of the target variable of interest is the main task and the other two adjustments may be, in some sense, regarded as signal extraction in a broad sense and in specific contexts. These adjustments have been extensively discussed in the literature on time-series analysis since the 1960s. Good reviews are available in Fernandez (1981), Cholette and Dagum (1994) and Hillmer and Trabelsi (1987).

Signal extraction is done to reduce the impact of noise in an observed time series where the signal evolves over time according to a stochastic process. The estimated signal is regarded as an estimate of the true series with fewer errors than in the original series. Signal extraction in the strict sense assumes an explicit statistical model for the signal and is usually done to process data from repeated surveys. When two or more series are available on the same variable, maybe with different frequencies, the more reliable measurements are taken as benchmarks and benchmarking is an exercise to make the other series consistent with the benchmarks. Benchmarking may also be treated as an exercise to combine optimally two or more series of measurements to achieve improved estimates of the signal.

Durbin and Quenneville (1997) have discussed the use of state-space models to benchmark time-series data. They consider monthly or quarterly data which they represent by state-space structural time-series models, and take annual totals as benchmarks. In a two-stage method for benchmarking, they combine information from the stage where they use state-space models with information from the benchmarks. In the single-stage method, they first combine the two series into one to which they apply the state-space model. They show that this approach is better than the ARMA model to represent the monthly data at the first stage. They consider both additive and multiplicative models and provide a comprehensive treatment of the subject.

10.4 Functional Data Analysis

Functional data analysis (FDA) refers to data generated and recorded at different points of time or for different wavelengths or, generally, against different values or levels of some entity such that data for any individual or unit over these values or levels (maybe even labels) are more or less similar and are definitely correlated. In fact, the data for any individual may be taken usually as the discretized version of a continuous curve. The objectives of the analysis could be to smooth these curves, to study some features like the derivatives of the first few orders, to distinguish one curve from another or to explain the pattern of the curve in relation to the values of the variable(s) to which points on the curve correspond. Successive points of time are generally taken to be equidistant and the number of time points is generally much more than the number of individuals or units for which the functional variable is noted. One of the earliest studies where FDA was used was in a study of height for ten girls growing over the first 20 years of life. More frequent measurements of height were taken in early ages, followed by annual measurements. Functional data analysis is not restricted to time-series data, though functional data quite often appear as multiple time series.

Four different situations where functional data analysis is appropriate were mentioned in Rao et al. (2008). These are:

1. We have a simple set of curves, one for each individual, and we want to find the mean curve or to examine the shapes of the curves in terms of their derivatives.

2. The different curves correspond to different labels or categories of a categorical study variable, e.g. groups of individuals, and we may be interested to distinguish curves belonging to one category from those belonging to another. It is also possible to note a univariate or multivariate study variable Y and our interest could lie in the regression of Y on the curves. The terms functional regression or signal regression have been used to denote this situation.

3. The study variable is also a functional variable and we may want to know how the two sets of curves influence each other.

4. Only the study variable Y is functional and the explanatory variables are non-functional, like factors affecting the functional variable. We then find an analogue of ANOVA.

In the second situation, the regression coefficients cannot be estimated by the ordinary least-squares method since the number of variables far exceeds the number of data points (units). One can use the principal-component-based partial least-squares approach or refer to functional principal-component analysis (FPCA). FPCA is an important dimension-reduction tool and in sparse data situations can be used to impute functional data. Penalized splines have also been used to estimate regression coefficients. Classic functional regression models are linear, though non-linear models have also been investigated. This type of non-parametric functional regression has been discussed in the literature (Ferraty and Vieu, 2006). In fact, functional regression is an active area of research and includes combinations of (1) functional responses with functional covariates, (2) vector responses with functional covariates and (3) functional responses with vector covariates.

Clustering and classification become more challenging in the context of functional data. In the terminology of machine learning, functional data clustering is an unsupervised learning process, while functional data classification is a supervised learning procedure.

10.5 Non-Parametric Methods

The classic periodogram introduced long ago by Schuster (1898) as the precursor of spectral analysis may be regarded as a non-parametric tool for time-series analysis. If the auto-correlation structure of a stationary process is of interest, the spectral density may be estimated as a summary of the second-moment properties. Since prediction is the end objective in any time-series analysis, we are interested in studying conditional means when a point prediction is required and conditional variances and even complicated conditional densities if interval forecasts or assessments of future volatility are needed.

To analyse the conditional mean in the non-parametric approach, we may start with the simple model $X_t = f(X_{t-1}, X_{t-2} ...) + \varepsilon_t$ where ε_t is a series of innovations which is independent of past X_t. In this model, $f(.)$ represents the conditional expectation at time t, given the past observations $X_{t-1}, X_{t-2}, ...$ and it is the minimum mean squared error one-step predictor for X_t.

Efron and Tibshirani (1993) discuss estimation of standard errors of linear autoregressive parameter estimates using the bootstrap approach. Sose evaluates the distribution of the parameter estimates of an AR (1) model by the bootstrap and this exercise has been extended to the case of the ARMA (p, q) process. The bootstrap method has been used for spectral estimation. It is also possible to apply bootstrap to the observations in a time series by sampling blocks of observations. Given a series $X_1, X_2, ... X_n$ all possible blocks of $1 <$ n observations are considered and random samples of such blocks are drawn and joined together to form a bootstrap time series of length roughly n. Repeating this process m times we get m bootstrap time series to be used for studying the distributional properties of the original time series.

The functional box-plot approach has been used by Ngo (2018), extending this approach to surfaces. The functional box plot produces a median curve – not equivalent to connecting medians obtained from situation-specific box plots real-life data. Surface box plots have been used to explore the variation of the spectral power for the alpha and beta frequency bands across the brain cortical surface. He also used a rank-based non-parametric test to investigate the stationarity of EEG traces across an examination acquired during resting state by comparing the early vs. late phases of a single resting-state EEG assessment.

Among the specific non-parametric tests used in time-series analysis, mention can be made of the Mann-Kendall trend test for the possible existence of a monotonic trend, even when the data reveal some seasonal pattern. The statistic for a univariate series is given by $S = \Sigma \Sigma \text{sgn}(X_j - X_k)$, the first sum over k from 1 to n – 1 and the second over j from k + 1 to n, sgn (x) = 1 if x > 0, = 0 if x = 0 and = −1 if x < 0. This is really speaking D τ, where τ is Kendall's tau and D depends on n and differences among time points. The Cox-Stuart trend test tests the first third of the series against the last third for trend.

Petitt's test can be used to determine whether there is a point at which the value(s) in the data change. The statistic used $K_T = \max | U_{t,T} |$, where $U_{t,T} = \Sigma \Sigma \text{sgn}(X_i - X_j)$, the first sum over I from 1 to t and the second over j from I + 1 to T. The change point in the series is located at K_T provided the statistic is significant. The significance probability of K_T is approximated for $p \leq 0.05$ with $p \approx 2 \exp(-6 K_T^2 / [T^2 + T^3])$.

10.6 Volatility Modelling

Volatility is a cause of worry for decision-makers and any satisfactory forecast of the future outcome of a current decision has to take due account of volatility. Measures and models for volatility became more visible in the context of time series of stock prices or asset values and thus, in financial market behaviour analysis, the concept of volatility is pretty general and applicable in the context of any variable(s) of interest whose value(s) is (are) likely to change with time in a manner that the change during a future interval of time cannot be exactly predicted.

Volatility has been simply defined as a measure of variability like the standard deviation or the coefficient of variation of change (during successive periods). In the case of a stationary time series when the expectation of change is taken as zero, standard deviation is the measure of volatility. Thus, if P_t, $t = 1, 2, \ldots T$ denotes the series on price and changes are measured by $\log P_t / P_{t-1} = R_t$ (something like the rate of return) then the SD σ of R_t defines volatility, usually with annual data, and then we can consider $\sqrt{12}\,\sigma$ as the monthly measure and $\sqrt{52}\,\sigma$ as the weekly measure, if we so need. Estimation of σ from sample data can be carried out in several possible ways, depending on the data we use and the average we choose. Thus we can use all the available data or only a recent part after some possible innovation. The average could be a historical average or a moving average or an exponentially weighted moving average. In fact, the overall variance versus the conditional variance is a big issue even in defining and interpreting volatility.

Volatility itself could be stochastic in nature. Stochastic volatility models are those in which the variance of a stochastic process is itself randomly varying. In fact, these models provide one approach to resolve a shortcoming of the famous Black-Scholes model, which assumes a constant volatility over the life of the derivative, unaffected by changes in the price level of the underlying security. These models start with constant volatility models and then treat this variance as a Brownian motion. Some of the stochastic volatility models are the Heston model, with a volatility proportional to the square root of its level, the constant elasticity of variance (CEV) model, which can accommodate a possible increase (decrease) in volatility with an increase (decrease) in prices, the stochastic alpha beta rho (SABR) model, the generalized autoregressive conditional heteroscedasticity (GARCH) model and its many variants, with volatility being a pre-determined constant at time t, given previous values, and the 3/2 model, which is similar to the Heston model. The GARCH model has been extended to the multivariate set-up as MGARCH to predict multivariate volatility.

11

Validation and Communication of Research Findings

11.1 Introduction

Results found and conclusions reached in a research study are not meant to be treasured secretly by the investigator(s), even when they contain confidential information that cannot be disclosed in the interest of national or international security. Research findings are generally meant for others or, at least, co-workers, collaborators, peers and in some cases the entire society. Because of the huge impact that such findings may have in some situations, it is essential that these findings and the manner in which they were arrived at be duly validated before being communicated to any one. And validation as a scientific exercise has to unequivocally establish the acceptability or otherwise of the research findings. In fact, validation itself may require collection and appropriate analysis of data (maybe mostly secondary) following a proper design that chooses the relevant criteria for validity which can be implemented with the available data.

Some aspects of validity are relatively easy to check. Thus internal inconsistencies, say, between objectives and research design or between nature and volume of data to be analysed and tools of analysis deployed, or between the results of analysis and the conclusions reached, or any deficiency in the experimental design or the survey design and the like can be detected by a careful perusal of the research findings that may be contained in a (draft) document for internal use. Sometimes, numerical computations may reveal errors or may even point out some gross discrepancies in numerical results when parameters involved in some result derived in the research (and not numerically verified) are assigned some realistic values.

Validation essentially implies some comparison with earlier research findings which have been duly validated, documented and communicated through appropriate media. Unlike in psychology, where reliability and validity are different concepts justifying different measures, reliability with the broad implication of reproducibility and bearing an analogy with test-retest reliability in psychometry is very much a component of validity in the present context.

11.2 Validity and Validation

Most researchers accept that the following types of validity should be assessed in respect of findings of the research which are meant to be valid beyond their limited contexts or even when the research findings are one of a 'one-off' type not meant for generalization, e.g. a highly interpretive case study. Some argue that in the latter case, validity should better relate to the research process, including methods to collect data, analyse data and interpret the results of analysis.

Construct validity: established through the correct design and use of data-collection tools for the specific concepts being studied. Correctness implies relevance and adequacy in relation to such concepts as are reflected in the research objectives and working hypotheses.

Internal validity: is important if any relationships among findings of the research are desired as an objective of research. This implies a situation where research involves different distinguishable aspects of the phenomenon under study on which relevant findings are being reported and relationships among such findings are expected.

External validity: is to be insisted upon whenever findings of any research are being claimed as generalizable to an extent – to be better indicated by the researcher. Such generalizability would mean that the findings will be similar if tried out at other locations on other occasions by other research workers.

Reliability: requires adoption of a credible and consistent line of enquiry and data collection to ensure that the same data would arise if the adopted line is repeated in a different setting.

Conceptual validity: suggests that the constructs under investigation and any philosophical assumptions made about those are duly reflected in the concerned steps of the research process.

Among different methods tried out to establish validity, triangulation has been widely accepted in the context of social science research. Triangulation makes use of different methodologies, from both positivist and interpretive epistemologies, and was developed by researchers who believed that deductive and inductive research are really not mutually opposed to each other, rather they focus on different dimensions of the same phenomenon and together provide a better comprehension of the phenomenon. Thus triangulation as a tool for validation may compare and combine evidence collected in different ways, like designing and conducting an experiment on the phenomenon and planning and carrying out an observational study or getting access to metadata and the like. For the same line of data collection, like conducting a response-based survey, triangulation may involve two different groups of investigators canvassing the same responding units. Different data sources can be tapped. For example, in a study of eating habits, initial coding of a transcript of group discussions among office workers may lead to the identification of two sub-categories, viz. leisure and work. This may lead the researcher to carry out a semi-structured interview with a professional cook to explore the relevance of the context, viz. what transpires in a group discussion with the experience of eating.

11.3 Communication of Research Findings

Findings or results of any research are meant to benefit others and hence should be properly communicated, usually through publication of papers or articles in journals or edited volumes which have a good circulation among persons likely to be interested in the findings – to enhance their knowledge or to carry out further studies on the phenomena under investigation. An early publication (maybe web-based) is desired.

Sometimes, researchers may find it useful to communicate their findings to competent peers or researchers in the same field to elicit their comments and suggestions, which should be acted upon to improve the quality of the findings and/or of the presentation. In fact, interaction with the peer group once results of some experiment or survey have come up through some form of analysis may help the research worker(s) in revisiting the results as well as the procedures followed. The apprehension that such results and the procedures in the background where a certain amount of originality is always expected may be misused by some in the peer group can be allayed in various ways.

Since research into a phenomenon or an aspect thereof does not generally end in one go, the findings of the research (as one natural round of research) should be communicated through various available media not merely among peers but also among those who are likely to be interested in the findings and even to a wider audience if the findings do or can find useful applications that may influence the thoughts and works of others. In the case of sponsored research, specially when a private agency has financially supported the research, some curbs could be there on the extent of communication of the research findings Some grant-giving institutions in government may require communication of the research findings to some mutually identified experts or referees. While the spirit behind this approach is to help the research team derive benefits from constructive comments and suggestions by the referees or experts for improving upon the substantive and/or technical content of the work and its presentation, seldom are such comments or suggestions received.

The findings of some empirical and even theoretical research specially those with some popular appeal are also reported in newspapers and magazines. The requirement by the media for a presentation that can be understood without much difficulty by a large cross-section of the readers or the audience sometimes boils down to diluting the rigour or the rigmarole usually associated with any meaningful research.

11.4 Preparing a Research Paper/Report

Preparing a research paper/article for publication in a journal deserves some attention.

Usually such an article should have the following parts. Some journals may specify the parts or the format for preparing the manuscript.

Title (sub-title, if needed to make the title not too long or obscure) – which should focus on the most important contribution of the author to the problem considered. It should not be anything not much discussed in the text.

Abstract – which should be a brief note on the problem taken up, the approach broadly followed and the findings. It should not give any details about the context.

Introduction – as distinct from the Abstract, should indicate the context and the problem under study along with the materials and methods used together with the results obtained. A brief review of relevant literature indicating gaps therein should form a part of the Introduction. An idea about the content of the paper in its various sections/subsections may also be added here.

Materials and Methods – which may include a section on symbols and notations used throughout the text and continue to give the problem statement, derivation of results (maybe by way of theorems and supporting lemmas), numerical illustrations wherever desired, etc. This being the major part of the paper, there should be sections and sub-sections with sub-headings, if necessary, for parts of the text which can be considered as distinct, though related, elements of the content.

Numerical Illustrations and Simulation Results along with their interpretations could form a separate part of the paper, in some cases.

Concluding Remarks, if added at the end, should not reproduce parts of the Abstract or the Introduction. It should rather indicate the merits and limitations of the work and, if possible, indicate aspects or issues related to the work which call for further investigation.

References – this should list in a prescribed manner all the publications mentioned in the text with the usual details about author(s), journal / book / edited volume, publishers, volume and issue number, page numbers etc. No entry in the References should be left out from the text and no entry referred to in the text should be missing from References.

A research report providing all relevant details about a research project undertaken by an individual or a team or even a whole institution – sometimes in collaboration with others – should be carefully prepared not just to satisfy certain mandatory requirements set forth by a grant-giving agency or by the institution administration itself but also for archival purposes to benefit the posterity of interested individuals. Such a report should provide all relevant details to comprehend the end results or finding,s like sample design, data collection procedures along with instruments and methods for measurements, data analysis and summary tables, conclusions reached, publications based on the research methods and findings, any feedback received and the like. Preparation of the report must facilitate archiving in some desired format. In the case of huge data having been collected, a storage device containing the data should be an important attachment to the main body of the report.

Acknowledgement becomes a necessity in some cases, e.g. where the study being reported was supported by some agency or in situations where existing protocols require the help and support received from superiors and peers to be acknowledged. A complete list of references used by the research team and required by the readers or users of the report is a must. It must be clearly noted that References and Bibliography are not synonymous. The latter is a list of all publications that are relevant to the work contained in the paper which a reader can go through for a comprehensive understanding of the subject, and not merely what is contained in the paper.

11.5 Points to Remember in Paper Preparation

A research paper meant for possible publication in some peer-reviewed journal is not the same as a document reporting the findings of a research activity. Some general principles should be followed in preparing such a paper.

These days different journals prescribe different formats for submission of papers. However, the general principles run through all these in some sense or another. And here is just an indicative (and not exhaustive) list of points to be kept in mind while preparing the manuscript of a research paper intended for publication in a refereed journal.

1. Title and sub-title, if any, must reflect the content of the paper and must not contain any aspect not dealt with in the text.

2. Abstract and Introduction are different. Introduction is to mention the problem, the context and the expected outcome (not the results) –a methodology outline may be added if it is somewhat or completely different from that used earlier. The abstract provides a concise statement of the findings and its importance or novelty or distinctiveness.

3. A critical review of previous work, indicating gaps therein, may be a part of the context and, thus, of the Introduction, if the paper is not lengthy. Otherwise, it can be taken up separately. Lengthy reviews, particularly those bordering on summaries of methods followed and results obtained in a previous study, should be avoided, unless where absolutely necessary to explain the content of the paper under preparation.

4. Data used should be completely comprehensible in terms of definitions of terms and phrases, sources and the manner of collection, effort to overcome problems associated with missing observations, suspect observations, changes in definitions or reference periods, etc.

   Reference periods of different data sets to be compatible or mutually consistent, e.g. NSSO employment data and ASI employment data.

5. Existing methods for data analysis need not be elaborated; instead adequate references may be provided. However, any modification or generalization attempted in the paper should be clearly stated in the necessary detail.

6. Results given in tables need not be presented through graphs and charts generally. Each table and diagram must be complete with a heading and descriptions of rows and columns. Footnotes may be added whenever necessary.

7. It must be remembered that sophistication in analysis does not compensate for any deficiency in the data analysed and does not necessarily add to the substantive content of the paper. It may contribute to the technical content, no doubt.

8. Non-standard terms and phrases and abbreviations have to be avoided even if some abbreviations are in common use. If a particular term or phrase which is relatively long occurs in many places in the body of the paper, abbreviated forms can be used only after the abbreviated form has been added, say in parentheses, after the fully expanded form for the first time. This is quite a necessity where seemingly identical abbreviations may have altogether different connotations, e.g. GT for 'grounded theory' and also for 'game theory'. Only standard definitions of technical terms must be used. Any deviation from the standard, if made, has to be justified.

9. Instruments used for any special measurement and the method of measurement should be mentioned in sufficient detail if these measurements are basic to the research objectives and the results thereof. Controls over the experimental environment whenever desired should also be indicated. These days, it has become quite desirable to provide an idea about the uncertainty in measurements of the important parameters dealt with.

10. Calculations of standard or routine statistical measures need not be shown, only the final results should be incorporated. Generally, original data sets are not to be provided, unless otherwise mandated and are relatively small in volume.

11. Use of statistical methods for making inductive inferences must be done carefully, keeping in mind conditions for applicability of certain statistical procedures, e.g. 'independently and identically distributed' set-up or normality of the underlying distribution, or homoscedasticity or symmetry or stationarity or infinitesimally small probability.

12. As pointed out in the previous section, References and Bibliography are not the same.

References include only those documents which have been referred to duly in the text, while Bibliography covers the entire relevant literature within the knowledge of the author. The general tendency to include cross-references under 'References', particularly those which are not easily accessible, should be avoided. A long list is not necessarily a mark of extensive knowledge of the author in the subject or theme presented in the paper.

References and Suggested Reading

Aly, E., and Behnkerouf, L. (2011) A new family of distributions based on probability generating functions. *Sankhya* B, 73, 70–80.

Allen, N.J. (1990) The measurement and antecedents of affective, continuance and normative commitment to the organisation. *Journal of Occupational Psychology*, 63, 1–18.

Amaratunga, D. (2002) Qualitative and quantitative research in the built environments. www.researchgate.net.

Anderson, T.W. (1984) *An Introduction to Multivariate Statistical Analysis* (2nd edn.). John Wiley, New York.

Andrews, D.F., and Herzberg, A.M. (1979) The robustness and optimality of response surface designs. *Journal of Statistical Planning and Inference*, 3, 249–257.

Arnold, B.C. (2004) A new method for adding a parameter to a family of distributions with applications to the exponential and Weibull families. *Biometrika*, 84(3), 641–652.

Averous, J., and Dottet-Bernadet, J.L. (2004) Dependence for Archimedean copulas and ageing properties of their generating functions. *Sankhya*, 66, part 4, 607–620.

Azzalini, A (1985) A class of distributions which includes the normal ones. *Scandinavian Journal of Statistics*,12, 171–178.

Banker, R.D. (1993) Maximum likelihood, consistency and data envelopment analysis: a statistical foundation. *Management Science*, 39(10), 1265–1273.

Banker, R.D., and Morey, R.C. (1986) The use of categorical variables in data envelopment analysis. *Management Science*, 32(12), 1613–1627.

Bannantine, J.A., Comer, J.J., and Handrock, J.L. (1989) *Fundamentals of Metal Fatigue Analysis*. Pearson, London.

Bar-Hen, A., and Daudin, J.J. (1995) Generalisation of the Mahalanobis distance in the mixed case. *Journal of Multivariate Analysis*, 5(1), 332–342.

Barlow, R.E., and Proschan, F. (1975) *Statistical Theory of Life Testing and Reliability*. Holt, Rinehart and Winston, New York.

Barmi, H. El, and Mukherjee, H. (2009) Peakedness and peakedness ordering in symmetric distributions. *Journal of Multivariate Analysis*, 100, 594–603.

Barnard, J., Rubin, D.B. and Zanutto, E. (1997) Lecture notes of the short course on multiple imputation for missing data, Utrecht, November 20–21.

Bartlett, M.S. (1937) Some examples of statistical methods of research in agriculture and applied botany. *Journal of the Royal Statistical Society: Series B (Statistical Methodology)*, 4, 137–170.

Basu, D.K. (1971) An essay on the logical foundation of survey sampling, Part I. *In Foundations of Statistical Inference* (Godambe and Sprott, eds.). Holt, Rinehart and Winston, Toronto.

Bawa, V.S. (1975) Optimal rules for ordering uncertain prospects. *Journal of Financial Economics*, 2, 95–121.

Benjamini, Y. (2010) Simultaneous and selective inference. Current successes and future challenges. *Biomedical Journal*. 52, 708–721.

Benjamini, Y., and Hochberg, Y. (1995) Controlling the false discovery rate: a practical and powerful approach to multiple testing. *Journal of the Royal Statistical Society: Series B (Statistical Methodology)*, 57, 289–300.

Benjamini, Y., and Yeutielei, D. (2001) The control of false discovery rate in multiple testing under dependency. *Annals of Statistics*, 29, 1165–1188.

Berger, J.O. (2003) Could Fisher, Jeffreys and Neyman agree on testing ? *Science*, 18(1), 1–32.

Bernardo, J.M., and Smith, A.F.M. (1996) *Bayesian Theory. Wiley Series in Probability and Statistics*. Wiley, Chichester.

Bernoulli, D. (1738) Hydrodynamique. *Opus Academicum*.

Bethel, J.W. (1989) Sample allocation in multivariate surveys. *Survey Methodology*, 15(1), 46–57.

Bhuyan, K.C. (2005) *Multivariate Analysis and its Applications.* New Central Book Agency, Calcutta.

Bicke, P.J., and Herzberg, A.M. (1979) Robustness of design against autocorrelation in time. Asymptotic theory, optimality for location and linear regression. *Annals of Statistics, 7,* 77–95.

Binder, C. (1996) Behavioral fluency: evolution of a new paradigm. *The Behavior Analyst,* 19, 163–197.

Birnbaum, Z.W. (1948) On random variables with comparable peakedness. *Annals of Mathematical Statistics,* 19, 76–81.

Biswas, A. (2001) Adaptive designs for binary treatment responses in phase III clinical trials. *Statistical Methods in Medical Research,* 10, 353–364.

Borg, I., and Gorenen, P.J.F. (2005) *Multidimensional Scaling* (2nd edn.). Springer, New York.

Boruch, R.F. (1971) Assuring confidentiality of responses in social research. *The American Sociologist,* 6(4), 308–311.

Bose, A. (1988) Edgeworth correction by bootstrap in autoregression. *Annals of Statistics,* 16, 1709–1722.

Box, G.E.P. (1949) A general distribution theory for a class of likelihood criteria. *Biometrika,* 36, 317–346.

Box, G.E.P. (1966) A note on augmented designs. *Technometrics,* 8(1), 184–188.

Box, G.E.P., and Behnken, D.W. (1960) Some new 3-level designs for study of quantitative variables. *Technometrics,* 2, 455–460.

Box, G.E.P., and Draper, N.R. (1975) Robust designs. *Biometrika,* 62, 347–352.

Box, G.E.P., and J.S. Hunter (1957) Multi-factor designs for exploring response surfaces. *Annals of Mathematical Statistics,* 28(1), 195–241.

Box, G.E.P., and Wilson K.B. (1951) On the experimental attainment of optimum conditions. *Journal of the Royal Statistical Society: Series B (Statistical Methodology),* 13(1), 1–45.

Breiman, L. (2001) Random forests. *Machine Learning,* 24, 95–122.

Breiman, L., Olshen, R.A., and Stone, C.J. (1984) *Classification and Regression Trees.* Wadsworth and Brooks/Cole, Basel.

Brewer, K.R.W. (1999) Design-based or prediction-based inference? Stratified random vs. stratified balanced sampling. *International Statistical Review,* 67(1), 35–47.

Brunk, H.D., et al. (1996) Estimation of the distributions of two stochastically ordered random variables. *Journal of the American Statistical Association,* 61, 1067–1080.

Bryant, A., and Charmaz, C. (2007) *The SAGE Handbook of Grounded Theory.* Sage, London.

Bryman, A. (2007) The research question in social research: what is its role? *International Journal of Social Research Methodology,* 10, 5–20.

Bryman, A., and E. Bell (2011) *Business Research Methods* (3rd edn.). Oxford University Press, Oxford.

Cai, Z., and Xu, X. (2002) Nonparametric quantile estimation for dynamic smooth coefficient models. Working paper, University of North Carolina, Charlotte.

Cantwell, J. (2008) Changing the modal context. *Theoria,* 74(4)b.

Chadwick, B.A., Bahar, H.M., and Albrecht, S.L. (1984) *Content Analysis in Social Science Research Methods* (Chadwick, B.A., et al., eds.), pp. 239–257. Prentice Hall, Upper Saddle River, NJ.

Chambers, R., and Tzavidis, N. (2008) M-quantile models for small area estimation. *Biometrika,* 73, 597–604.

Chandra, A., and Mukherjee, S.P. (2017) EOQ and its estimation in a static one-point inventory model. *IAPQR Transactions* 42(1), 65–75.

Charmaz, C. (2006) *Constructing Grounded Theory: A Practical Guide through Qualitative Analysis.* Sage, London.

Charnes, A., Cooper, W.W., and Rhodes, E. (1978) Measuring the efficiency of decision-making units. *European Journal of Operational Research,* 2(6), 429–444.

Charnes, A., Cooper, W.W., Lewin, A.Y., and Seiford, L.H. (1994) *Data Envelopment Analysis: Theory, Methodology and Applications.* Kluwer Academic, Dordrecht.

Chatterjee, S. (1968) Multivariate stratified surveys. *Journal of the American Statistical Association,* 63, 530–534.

Chaudhuri, A. (2010) *Essentials of Survey Sampling.* Prentice Hall of India, New Delhi.

Chaudhuri, A. (2014) *Modern Survey Sampling.* CRC Press, Boca Raton, FL.

Chaudhuri, A. (2018) *Survey Sampling*. Taylor & Francis, Boca Raton, FL.

Chaudhuri, A., and Dutta, T. (2018) *Determining the size of a sample to take from a finite*

Chaudhuri, P., et al. (1994) Piece-wise polynomial regression trees. *Statistica Sinica*, 4, 166–182.

Chaudhuri, P., et al. (1995) Generalised regression trees. *Statistica Sinica*, 5, 641–666.

Cheng, B., and Titterington, D.M. (1994) Neural networks: a review from a statistical perspective. *Statistical Science*, 9(1), 2–54.

Cherchye, L., and Post, T. (2003) Methodological advances in DEA: a survey and an application in the Dutch electricity sector. *Statistica Neerlandica*, 57(4), 410–438.

Cholette, P.A., and Dagum, E.B. (1994) Benchmarking time series with auto-correlated survey errors. *International Statitical Review*, 62(3), 365–377.

Clyde, M. and George, E.I. (2004) Model uncertainty. Statistical Science, 19 (1), 81–94.

Clyde, M., DeSimone, H., and Parmigiani, G. (1996) Prediction via orthogonalized model-mixing. *Journal of the American Statistical Association*, 91, 1197–1208.

Cochran, W.G. (1953) *Sampling Techniques*. Wiley, New York.

Cohen, J., and Cohen, P. (1983) *Applied Multiple Regression / Correlation Analysis for the Behavioral Sciences*. Lawrence Earlbaum, Hillsdale, NJ.

Condra, L.W. (1993) *Reliability Improvements with Design of Experiments*. Marcel Dekker, New York.

Cooper, R.D., Schilder, P.S., and Sharma, J.K. (2012) *Business Research Methods*. McGraw Hill (Education), New Dilhi.

Cornell, J.A. (2002) *Experiments with Mixtures*. Wiley, New York.

Cox, D.R. (1972) Regression models and life tables (with discussion). *Journal of the Royal Statistical Society: Series B (Statistical Methodology)*, 34, 187–220.

Cox, T.F., and Cox, M.A.A. (2001) *Multidimensional Scaling*. Chapman Hall, London.

Crick, F. (1994) *The Astonishing Hypothesis*. Simon & Schuster, New York.

Cristianinin, N., and Shawe-Taylor, J. (2000) *An Introduction to Support Vector Machines*. Cambridge University Press, Cambridge.

Crowder, M. (1989) A multivariate distribution with Weibull connection. *Journal of the Royal Statistical Society: Series B (Statistical Methodology)*, 51, 93–107.

Dagum, C. (1996) A systematic approach to the generation of income distribution models. *Journal of Income Distributions*, 6, 105–226.

Dale, C.J. (1986) Applications of the proportional hazard model in the reliability field. *Reliability Engineering*, 10, 1–14.

Deemer, W.L. Jr., and Votaw, D.F. Jr. (1955) *Annals of Mathematical Statistics*, 26, 498.

Dekhale, J.S., Prasad, R., and Gupta, V.K. (2003) Analysis of inter-cropping experiments using experiments with mixtures methodology. *Journal of the Indian Society of Agricultural Statistics*, 56, 260–266.

Deming W.E. (1950) *Some Theory of Sampling*. Wiley, New York.

Denison, D.G.T., et al. (1998) A Bayesian CART algorithm (with comment). *Biometrika*, 85, 363–377.

Derringer, G., and Suich, R. (1980) Simultaneous optimisation of several response variables. *Journal of Quality Technology*, 12, 214–219.

Deville, J.C. (1991) A theory of quota surveys. *Survey Methodology*, 17(2), 163–181.

Diaz-Garcia, J.A., and Cortez, L.U. (2008) Multi-objective optimisation for optimum allocation in multivariate stratified sampling. *Survey Methodology*, 34, 215–222.

Dickhaus, T., and Stange, J. (2013) Multiple point hypothesis testing problem and effective number of tests for control of the family-wise error rate. *Calcutta Statistical Association Bulletin*, 65, 123–144.

Dickhaus, T., et al. (2012) How to analyze many contingency tables simultaneously in genetic association studies. *Statistical Applications in Genetics and Molecular Biology*, 11, Article 12.

Dimitrov, S. (2014) Comparing data envelopment analysis and human decision making unit rankings: a survey approach. *Economic Quality Control*, 29(2), 129–141.

Draper, D. (1995) Inference and hierarchical modelling in the social sciences. *Journal of Educational and Behavioral Statistics*, 20(2), 115–147.

Durbin, J., and Quenneville, B. (1997) Benchmarking by state space models. *International Statistical Review*, 65(1), 23–48.

Dykstra, O. (1971) The augmentation of experimental data to maximize | X'X |. *Technometrics*, 13(3), 682–688.

Dyson, R.G., and Thanassoulis, E. (1988) Reducing flexibility in data envelopment analysis. *Journal of the Operational Research Society*, 39(6), 573–576.

Efron, B., and Tibshirani, R.J. (1993) *An Introduction to the Bootstrap*. Chapman and Hall, New York

Everitt, B.S. (1974) *Cluster Analysis*. Heinemann Educational Books, London.

Everitt, B.S., and Rabe-Hesketh (1997) *The Analysis of Proximity Data*. Arnold, London.

Farlie, D.J.G. (1960) The performance of some correlation coefficients in a general bivariate distribution. *Biometrika*, 47, 307–323.

Fay, B. (1996) *Contemporary Philosophy of Social Science: A Multi-Cultural Approach*. Blackwell, Cambridge.

Fedorov, V.V. (1972) *Theory of Optimal Experiments*. Academic Press, New York.

Fernandez, R. (1981) A methodological note on the estimation of time series. *Review of Economics and Statistics*, 63(3), 471–476.

Ferraty, F., and Vieu, P (2006) *Non-Parametric Functional Data Analysis*. Springer, New York.

Feynman, R.P. (1992) *Surely You're Joking Feynman*. W.W. Norton, New York.

Finner, H., Dickhaus, T., and Roters, M. (2007) Dependency and false discovery rate: asymptotics. *Annals of Statistics*, 35, 1432–1455.

Fisher, R.A. (1936) The use of multiple measurements in taxonomic problems. *Annals of Eugenics*, 7(2), 179–188.

Folks, J.L., and Antle, C.E. (1965) Optimum allocation of sampling units to strata when there are R responses of interest. *Journal of the American Statistical Association*, 60(309), 225–233.

Freedman, P. (1960) *The Principles of Scientific Research*. Pergamon Press, New York.

Fryzlewicz, P. (2014) Wild binary segmentation for multiple change point detection. *Annals of Statistics*, 42(6), 2243–2281.

Geisser, S. (1993) *Predictive Inference—An Introduction*. Chapman and Hall, New York.

Gelfand, A.E., Dey, D., and Chang, H. (1992) Treatment of missing survey data. *Survey Methodology*, 12, 1–16.

Ghosh, B.N. (1982) *Scientific Methods and Social Research*. Sterling, New Delhi.

Ghosh, S. (1990) *Statistical Design and Analysis of Experiments*. Marcel Dekker, New York.

Gill, R.D. (1984) Understanding Cox's regression model: a martingale approach. *Journal of the American Statistical Association*, 79, 441–447.

Glaser, B.G. (1998) *Advances in the Methodology of Grounded Theory*. Sociology Press, Mill Valley, CA.

Glaser, B.G. and Strauss, A.L. (1967) The discovery of grounded theory. Aldine Transactions, New Brunswick, NJ.

Glaz, J. (2000) *Probability Inequalities for Multivariate Distributions with Applications to Statistics*. Chapman and Hall / CRC Press.

Goode, W.J., and Hart, P.K. (1952) *Methods in Social Research*. McGraw Hill, New York.

Goodman, L.A. (1961) Snowball sampling. *Annals of Mathematical Statistics*, 32(1), 148–170.

Goswami, D., and S. Paul (2018) The decline in the aggregate productivity growth of Indian manufacturing. Evidence from plant-level panel data set. Paper presented in the 9th Seminar on Industrial Statistics, Calcutta.

Greenberg, B.G., Abul-Ela, A.A., Simmons, W.R., and Horvitz, D.G. (1969) The unrelated question randomised response model: theoretical framework. *Journal of the American Statistical Association*, 64, 520–539.

Hald, A. (1984) Nicholas Bernoulli's theorem. *International Statistical Review*, 52(1), 93–99.

Hansen, M.H., et al. (1953) *Sample Survey Methods and Theory*, Vol. I. Wiley, New York.

Hardle, W., and Vieu, P. (1992) Kernel regression smoothing of time series. *Journal of Time Series Analysis*, 13, 209–232.

Hardle, W., Lutkepol, H., and Chen, R. (1997) A review of non-parametric time series analysis. *International Statistical Review*, 65(1), 49–72.

Harman, H.H. (1976) *Modern Factor Analysis* (3rd edn.). Chicago University Press, Chicago.

Harrington, E.C. Jr. (1965) The desirability function. *Industrial Quality Control*, 21, 494–498.

Harris, T.E. (1948) Branching processes. *Annals of Mathematical Statistics,* 19(4), 474–494.

Hart, J.D. (1996) Some automated methods of smoothing time-dependent data. *Journal of Non-Parametric Statistics,* 6, 115–142.

Hartigan, J.A. (1975) *Clustering Algorithms.* John Wiley, New York.

Harvey, A.C., and Shephard, N. (1993) Structural time series models. In *Handbook of Statistics,* Vol. 11, pp. 261–302. Elsevier, Amsterdam.

Hebble, T.L., and Mitchell, T.J. (1972) Repairing response surface designs. Technometrics,14(3), 767–779.

Hedayat, A.S. (1981) Study of optimality criteria in design of experiments. In *Statistics and Related Topics* (Csorgo, M., et al., eds.). North Holland, Amsterdam.

Hedayat, A.S., and John, P.W.M. (1974) Resistant and susceptible BIB designs. Annals of Statistics, 2(1), 148–158.

Henley, E.J., and Kumamoto, H (1981) *Reliability Engineering and Risk Assessment.* Prentice-Hall, Upper Saddle River, NJ.

Herzberg, A.M. (1982) The robust design of experiments: a review. *Serdica,* 8, 223–228.

Hertzberg, A.M., and Andrews, D.F. (1978) The robustness of trans-block designs and cost-of-mail designs. Commun. Statistics, A, &, 479–85.

Hill, R.M. (1997) Applying Bayesian methodology with a uniform prior to the single-period inventory model. *European Journal of Operational Research,* 98(3), 555–562.

Hillmer, S.C., and Trabelsi, A. (1987) Benchmarking of economic time series. *Journal of the American Statistical Association,* 82, 1064–1071.

Hillway, T. (1964) *Introduction to Research* (2nd edn.). Houghton Mifflin, Boston, MA.

Hochberg, Y., and Benjamini, Y. (1990) More powerful procedures for multiple significance testing. *Statistics in Medicine,* 9, 811–818.

Hochberg, Y., and Tamhane, A. (1987) *Multiple Comparison Procedures.* Wiley, New York.

Holmberg, A. (2002) A multiparameter perspective on the choice of sampling design in surveys. *Statistics in Transition,* 5, 969–994.

Hommel, G. (1988) Stagewise rejective multiple test procedure based on a modified Bonferroni test. *Biometrika,* 75(2), 383–386.

Honda, T. (2004) Quantile regression in varying coefficient models. *Journal of Statistical Planning and Inference,* 121, 113–125.

Hougaard, P. (1986) A class of multivariate failure rate distributions. Biometrika, 73(3), 671–678.

Huberty, C.J. (1994) *Applied Discriminant Analysis.* John Wiley, New York.

Hwang, C.L., and Yoon, K. (1981) *Multiple Attribute Decision-Making: Methods and Applications.* Springer, New York.

Iawamura, K., and Liu, B. (1998) Chance-constrained integral programming models for capital budgeting in fuzzy environments. *Journal of Operational Research Quarterly,* 49(8) 854–860.

Jaynes, E.T. (1988) The relation of Bayesian and Maximum entropy method. *Maximum Entropy and Bayesian Methods in Science and Engineering,* 1, 25–29.

Jick, T.D. (1979) Mixing qualitative and quantitative methods: triangulation in action. *Administrative Science Quarterly,* 24, 602–611.

John, R.C.S., and Draper, N.R. (1975) D-Optimality for regression designs: a review. *Technometrics,* 17(1), 15–23.

Johnson, A.L., and Kuosmanen, T. (2012) One-stage and two-stage DEA estimation of the effects of contextual variables. *European Journal of Operational Research,* 220(2), 559–570.

Johnson, N.L., and Kotz, S. (1975) A vector multivariate hazard rate. *Journal of Multivariate Analysis,* 5(1), 53–66.

Johnson, R.A., and Wicher, D.W. (1996) *Applied Multivariate Statistical Analysis.* Prentice Hall of India.

Jose, K.K. and Paul, A. (2018) Marshall-Olkin Exponential Power Distribution and Its Generalization: Theory and Applications. *IAPQR Transactions,* 43(1), 1–30.

Kaiser, H.F. (1958) The varimax rotation for analytic rotation in factor analysis. *Psychometrika,* 23, 187–200.

Kalton, G. (1983) *Introduction to Survey Sampling.* Sage, London.

Kalton, G., and Kasprzyk, D. (1986) Model determination using predictive distributions with implementation via sampling-based methods. Technical Report, Stanford University, CA.

Kaplansky, I. (1945) A common error concerning kurtosis. *Journal of the American Statistical Association*, 40, 259.

Kardaun, O.J.W.F., et al. (2003) Reflections on fourteen cryptic issues concerning the nature of statistical inference. *International Statistical Review*, 71, 277–318.

Khan, M.G.M., Khan, E.A., and Ahsan, M.J. (2003) An optimal multivariate stratified sampling design using dynamic programming. *Australian & New Zealand Journal of Statistics*, 45, 107–113.

Khuri, A.I., and Conlon, M. (1981) Simultaneous optimisation of multiple responses represented by polynomial regression functions. *Technometrics*, 23. 363–375.

Kiaer, A. (1897) *The Representative Method of Statistical Surveys*. Central Bureau of Statistics, Norway.

Kiefer J. (1975) Construction and optimality of generalized Youden designs. In *A Survey of Statistical Designs and Linear Models*. (Stivastava, J.N., ed.). North Holland, Amsterdam.

Kim, M. (2007) Quantile regression with varying coefficients. *Annals of Statistics*, 35(1), 92–108.

Kish, L. (1965) *Survey Sampling*. Wiley, New York.

Klir, G.J., and Folger, T.A. ((1993) *Fuzzy Sets, Uncertainty and Information*. Prentice Hall of India, New Delhi.

Knight, J. (2008) Higher education in turmoil: the changing world of internationalisation. Google scholar.

Knight, S., Halkett, K.B. and Cross, D. (2010) The context and contextual constructs of research. 5th Conference on Qualitative Research in I.T., Nov. 29–30, Brisbane.

Koenker, R. (2005) *Quantile Regression*. Cambridge University Press, Cambridge.

Koenker, R., and Hallock, K. (2001) Quantile regression. *Journal of Economic Perspectives*, 15, 143–156.

Kokan, A.R., and Khan, S.U. (1967) Optimal allocation in multivariate surveys. *Journal of the Royal Statistical Society: Series B (Statistical Methodology)*, 29, 115–125.

Kothari, C.R. (2004) *Research Methodology*. Holt, Reinhart and Winston, New York.

Kozak, M. (2006) On sample allocation in multivariate surveys. *Communications in Statistics – Simulation and Computation*, 35, 901–910.

Kozak, M. (2006) Multivariate sample allocation: application of random search method. *Statistics in Transition*, 7, 889–900.

Krippendorff, K. (1980) *Content Analysis: An Introduction to its Methodology*. Sage, London.

Kruskal, J.B. (1964) Multi-dimensional scaling by optimising goodness of fit to a non-parametric hypothesis. *Psychometrika* 29(1), 1–27.

Kruskal, J.B., and Wish, M. (1978) *Multi-Dimensional Scaling*. Series in quantitative applications in the social sciences, 11. Sage, London.

Kshirsagar, A.M. (1972) *Multivariate Analysis*. Marcel Dekker, New York.

Kuk, A.Y.C. (1990) Asking sensitive questions indirectly. *Biometrika*, 77, 436–438.

Kunsch, H.R. (1989) The jackknife and the bootstrap for general stationary observations. *Annals of Statistics*, 17, 1217–1241.

Lachenbruch, P.A. (1975) *Discriminant Analysis*. Hafner Press, New York.

Landis, J.R., and Koch, G.G. (1977) The measurement of observer agreement for categorical data. *Biometrics*, 33(1), 159–174.

Lawley, D.N., and Maxwell, A.E. (1071) *Factor Analysis in Statistical Methods*. American Elsevier, New York.

Lawson, J., and Erjavec, J. (2001) *Modern Statistics for Engineering and Quality Improvement*. Duxbury, Pacific Grove, CA.

Leamer, E.E. (1978) *Specification Searches: Ad Hoc Inference with Non-Experimental Data*. John Wiley, New York.

Lewis, H.R., and Papadimitriou, C.H. (1999) *Elements of the Theory of Computation*. Prentice Hall of India, New Delhi.

Liski, E.P., Mandal, N.K., Shah, K.R., and Sinha, B.K. (1992) *Topics in Optimal Designs*. Lecture Series in Statistics, 163. Springer, New York.

Liski, E.P., Luoma, A., and Zaigraev, A. (1999) Distance optimality design criterion in linear models. *Metrika*, 49, 193–211.

Liski, E.P., and Zaigraev, A. (2001) A stochastic characterisation of Lowener optimality design criteria in linear models. *Metrika*, 53, 207–222.

Little, R.J.A. (1988a) Missing data in large surveys. *Journal of Business Economics & Statistics*, 6, 287–301.

Little, R.J.A. (1988b) A test of missing completely at random for multivariate data with missing values. *Journal of the American Statistical Association*, 83(404), 1198–1202.

Little, R.J.A., and Rubin, D.B. (1989) The analysis of social science data with missing values. Sociological Methods and Research, 18(2-3), 292–326.

Little, R.J.A., and Rubin, D.B. (2002) *Statistical Analysis with Missing Data*. Wiley, New York.

Liu, R.Y., and Singh, K. (1992) Moving blocks jackknife and bootstrap capture weak dependence. In *Exploring the Limits of Bootstrap* (Lepage, R., and Billard, L., eds.), pp. 225–248. Wiley, New York.

Locke, K.D. (2001) *Grounded Theory in Management Research*. Sage, London.

Loh, W.Y. (2002) Regression trees with unbiased variable selection and interaction detection. *Statistica Sinica*, 12, 361–386.

Machado, J.A., and Silva, J.S. (2000) Glejser's test revisited. *Econometrics*, 97(1), 189–202.

Mackellar, D.A. (2007) Perceptions of lifetime risk and actual risk. AIDS Behaviour, 11(2), 263–270.

Makarov, L.M., Vinogradskaya, T.M., Rubchinsky, A.A., and Sokolov, V.B. (1987) *The Theory of Choice and Decision-Making*. Mir, Moscow.

Mandal, N.K., Shah, K.R., and Sinha, B.K. (2000) Comparison of test vs. control treatments using distance optimality criterion. *Metrika*, 52, 147–162.

Marshall, A.W. (1975) Multivariate distributions with monotone hazard rate. In Fault Tree Analysis, SIAM (IAEA Conference), USA.

Marx, B.D., and Eilers, P.H.C. (1999) Generalised linear regression on sampled signals and curves, a p-spline approach. *Technometrics*, 41(3) 1–13.

Mee, R. (2002) Three-level simplex designs and their use in sequential experimentation. *Journal of Quality Technology*, 34(2), 152–164.

Minsky, M. (2003) What comes after minds? In *The New Humanists—Science at the Edge* (Brockman, J., ed.). Barnes and Noble, New York.

Mishra, P. (2015) *Business Research Methods*. Oxford University Press, Oxford.

Montgomery, D.C. (2012) *Design and Analysis of Experiments* (8th edn.). John Wiley, Chichester.

Morita, H, and Seiford, L.M. (1999) Characteristics on stochastic DEA efficiency—reliability and probability being efficient. *Journal of the Operations Research Society of Japan*, 42(4), 389–406.

Morse, J.M., et al. (2009) *Developing Grounded Theory: The Second Generation*. CRC Press, Boca Raton, FL.

Mukherjee, R. (1980) A generalized procedure for product scaling. *IAPQR Transactions*, 2, 71–83.

Mukherjee, S.P. (1982) Efficiency comparisons among estimators of reliability. *Calcutta Statistical Association Bulletin*, 185–194.

Mukherjee, S.P. (2012) Multi-response and non-traditional experiments. *Journal of the Indian Society of Agricultural Statistics*, 1–8.

Mukherjee, S.P. (2018) *Quality: Domains and Dimensions*. Springer, Singapore.

Mukherjee, S.P., and Chandra, A. (2011) Estimation of optimal order quantity in inventory models. *IAPQR Transactions*, 36(1), 27–44, 42(1), 65–74.

Mukherjee, S.P., and Mandal, A. (1994) Secretary selection problem with a chance constraint. *Journal of the Indian Statistical Association*, 32, 29–34.

Mukherjee, S.P. and Pal, M. (2017) Reliability improvement through designed experiments. *Journal of Applied Probability and Statistics*, 12 (1), 1–9.

Mukherjee, S.P., and Roy, D. (1986) Some characterisations of the exponential and related life distributions. *Calcutta Statistical Association Bulletin*, 35, 189–197.

Mukherjee, S.P., et al. (2018) *Statistical Methods in Social Science Research*, pp. 29–37, 47–50. Springer, Singapore.

Nair, V.N. (1992) Taguchi's parameter design: a panel testing methods for multivariate time series. Ph.D. dissertation. UC Irvine.

O'Hagan, A., and Leonard, T. (1976) Bayes estimation subject to uncertainty about parameter constraints. *Biometrika*, 63, 201–202.

Olesen, O.B., and Petersen, N.C. (2003) Identification and use of efficient faces and facets in DEA. *Journal of Productivity Analysis*, 20(3), 323–360.

Olson, C.L. (1974) Comparative robustness of six tests in multivariate analysis of variance. *Journal of the American Statistical Association*, 69, 894–907.

Olson, C.L. (1976) On choosing a test statistic in multivariate analysis of variance. *Psychological Bulletin*, 83, 579–586.

Orbe, F., and Rodriguez, P. (2005) Non-parametric estimation of time varying parameterts under shape restrictions. *Journal of Econometrics*, 126, 53–77.

Osgood, C., May, W., and Miron, M. (1975) *Cross-Cultural Universals of Affective Meaning*. Univ. of Illinois Press, Champaign.

Panneerselvam, R. (2004) *Research Methodology*. Prentice Hall of India, New Delhi.

Paris, P.C., and Erdogan, F. (1963) A critical analysis of crack propagation laws. *Journal of Basic Engineering*, 85, 528–543.

Park, D.C., et al. (1991) Electrical load forecasting using an artificial neural network. *IEEE Transactions on Power Engineering*, 6, 442–449.

Parkan, C. (1994) Operational competitiveness ratings of production units. *Managerial and Decision Economics*, 15, 201–221.

Petitt, A.N. (1979) A non-parametric approach to the change point problem. *Journal of the Royal Statistical Society, Series C (Applied Statistics)*, 28, 126–135.

Pignatielo, J.J. Jr. (1993) Strategies for robust multi-response quality engineering. *IIE Transactions*, 25, 5–15.

Pillai, K.C.S., and Gupta, A.K. (1969) On the exact distribution of Wilk's criterion. *Biometrika*, 56, 109–118.

Plackett, R.L., and Burman, J.P. (1946) The design of optimum multi-factor experiments. *Biometrika*, 33, 305.

Powell, D. (2016) Quantile regression with non-additive fixed effects. *Quantile Treatment Effects*. Available at: http://works.bepress.com/david_powell/1/.

Prendergast, J., Murphy, E., and Stephenson, N. (1996) Building in reliability –implementation and benefits. *International Journal of Quality & Reliability Management*, 13(5), 77–90.

Prescott, P., and Draper, N.R. (1998) Mixture designs for constrained components in orthogonal blocks. *Journal of Quality Technology*, 36, 413–431.

Pukelsheim, F. (1993) *Optimal Design of Experiments*. John Wiley, New York.

Puri, P.S., and Rubin, H. (1974) On a characterisation of the family of distributions with constant multivariate failure rates. *Annals of Probability*, 2(4), 738–740.

Qiu, P., Chi, A., and Li, X. (1991) Estimation of jump regression function. *Bulletin of Informatics and Cybernetics*, 24, 197–212.

Raftery, A.E., Madigan, D., and Volinsky, C.T. (1996) Bayesian model averaging for linear regression models. *Journal of the American Statistical Association*, 92, 1197–1208.

Raftery, A.E., Tanner, A.E., and Wells, M.T. (2002) *Statistics in the 21st Century*. Chapman and Hall, London.

Ramsay, J., and Silverman, B.W. (2002) *Applied Functional Data Analysis. Methods and Case Studies*. Springer, New York.

Ramsay, J., and Silverman, B.W. (2005) *Functional Data Analysis* (2nd edn.). Springer, New York.

Rao, C.R., and Shanbhag, D.N. (1998) 8 Recent approaches to characterizations based on order statistics and record values. *Handbook of Statistics*, 16, 231–256.

Rao, J.N.K. (2003) *Small Area Estimation*. Wiley, New York.

Rao, C.R., Shalabh, Toutenberg, H., and Heumann, C. (2008) *Linear Models and Generalizations*. Springer, Berlin.

ReliaSoft. (2014) *Reliability Engineering Resources*, Issue 155.

Retzlaff-Roberts, D., and Morey, C. (1993) A goal programming method of stochastic allocative data envelopment analysis. *European Journal of Operational Research*, 71, 379–397.

Ribeiro, J.L., and Elsayed, E.A. (1995) A case study on process optimisation using the gradient loss function. *International Journal of Production Research*, 33, 3233–3248.

Rosenberg, A. (1993) Hume and the philosophy of science. In *The Cambridge Companion to Hume* (Norton, D., ed.). Cambridge University Press, New York.

Rosenberger, W.E. (1996) New directions in adaptive designs. *Statistical Science*, 11, 137–149.

Royall, R.M. (1970) On finite population sampling under certain regression models. *Biometrika*, 57, 377–387.

Royall, R.M., and Hersen, J. (1973) Robust estimation in finite populations. *Journal of the American Statistical Association*, 68, 880–889.

Rubin, D.B. (1976) Inference and missing data. *Biometrika*, 63, 581–592.

Rubin, D.B. (1987) *Multiple Imputation for Non-Response in Sample Surveys*. Wiley, New York.

Rubin, D.B. (1996) Multiple imputation after 18+ years. *Journal of the American Statistical Association*, 91(434), 473–489.

Rumelhart, D.E., Hinton, G.E., and Williams, R.J. (1986) Learning internal representation by back-propagating errors. *Nature*, 323, 533–536.

Rummel, R.J. (1970) *Applied Factor Analysis*. Northwestern University Press, Evanston, IL.

Sahoo, B.K. (2015) Scale elasticity in non-parametric DEA approach. In Data Envelopment Analysis (Zhu, J., ed.), Chapter 9. Springer, New York.

Sarkar, S. (1998) Some probability inequalities for ordered MTP2 random variables : a proof of the Simes conjecture. Annals of Statistics, 26(2), 494–504.

Schafer, J.L. (1997) *Analysis of Incomplete Multivariate Data*. Chapman & Hall, New York.

Schlaifer, R. (1959) *Probability and Statistics for Business Decisions (3rd edn.)*. McGraw Hill, New York.

Schölkopf, B., and Smola, A. (2002) Support vector machines and kernel algorithms. Available from Semantic Scholar.

Schuster, A. (1898) On the investigation of hidden periodicities with applications to a supposed 26-day period of meteorological phenomena. *Journal of Geo-Physical Research*, 3(1), 13–41.

Sengupta, J.K. (1987) Data envelopment analysis for efficiency measurement in the stochastic case. *Computers & Operations Research*, 14, 17–169.

Shah, K.R., and Sinha, B.K. (1989) *Theory of Optimal Designs*. Lecture notes in Statistics, Series 54. Springer, New York.

Shaked, M., and Shanthikumar, J. (1994) *Stochastic Orders and their Applications*. Academic Press, New York.

Sharma, S. (2018) Corruption in public procurement. Unpublished dissertation in Jadavpur University.

Simes, R.J. (1986) Improved Bonferroni procedure for multiple tests of significance. *Biometrika*, 73(3), 751–754.

Sinha, B.K. (1970) A Bayesian approach to optimum allocation in regression problems. Calcutta Statistical Association Bulletin, 19(1), 45–52.

Sinha, B.K. (2002) Glimpses of design optimality criteria: Recent advances. Presidential address in the Statistics section of the Indian science Congress 89th session.

Sinha, B.K., et al. (2014) *Optimal Mixture Experiments*. Springer, Berlin.

Sinha, B.K., et al. (2016) *Optimal Covariate Designs and their Applications*. Springer, Berlin.

Sinha, B.K., Das, P., Mandal, N.K., and Pal, M. (2010) Parameter estimation in linear and quadratic mixture models. *Pakistan Journal of Statistics*, 26, 77–96.

Sinha, S.K., and Rao, J.N.K. (2009) Robust small area estimation. *Canadian Journal of Statistics*, 37, 381–399.

Smith, C.A.B. (1947) Some examples of discrimination. *Annals of Eugenics*, 15, 272–282.

Snee, R.D., and Marquart, D.W. (1974) Extreme vertices designs for linear mixture models. *Technometrics*, 16, 399–408.

Snee, R.D., and Marquart, D.W. (1976) Screening concepts and designs for experiments with mixtures. *Technometrics*, 18, 19–29.

Soric, B. (1989) Statistical discoveries and effect-size estimation. *Journal of the American Statistical Association*, 84(406), 608–610.

Stewart, I. (2002) The mathematics of 2050. In *The Next Fifty Years. Science In The First Half Of The Twenty-First Century* (Brockman, J., ed.). Vintage, New York.

Strauss, A.L., and Corbin, J. (1998) *Basics of Qualitative Research: Grounded Theory Procedures and Techniques* (2nd edn.). Sage, London.

Sudman, S. (1967) *Reducing the Cost of Surveys*. Aldaine, Chicago.

Taguchi, G., Elsayed, E.A., and Tsiang, T. (1989) *Quality Engineering in Production Systems*. McGraw Hill, New York.

Thompson, E., and Cosmelli, D. (2006) Neurophenomenology: an introduction to neurophilosophers. In *Cognition and the Brain: The Philosophy and Neuroscience Movement* (Brook, A., and Akins, K., eds.). Cambridge University Press, New York.

Thomson, I., and Holmoy, A.M.K. (1998) Combining data from survey and administrative record systems-the Norwegian experience. *International Statistical Review*, 66(2), 201–221.

Thurstone, L.L. (1927) A law of comparative judgement. *Psychological Review*, 34, 273–286.

Tone, K. (2000) A slacks-based measure of efficiency in data envelopment analysis. *European Journal of Operational Research*, 130, 498–509.

Tone, K. (2017) *Advances in DEA theory and Applications with Extensions to Forecasting Models*. John Wiley, Chichester.

Tong, H. (1990) *Non-linear Time Series Analysis: A Dynamic Approach*. Oxford University Press, Oxford.

Tsiatis, A.A. (1981) A large sample study of Cox's regression model. *Annals of Statistics*, 9, 93–108.

Wang, D., and Bakhai, A. (eds.) (2006) *Clinical Trials: A Practical Guide to Design, Analysis and Reporting*. Remedica, London.

Warner, S.L. (1965) Randomised response: a survey technique for eliminating evasive answer bias. *Journal of the American Statistical Association*, 60(309), 63–69.

Weber, R.P. (1990) *Basic Content Analysis*. Sage, Thousand Oaks, CA.

Williams, W.H., and Mallows, C.L. (1970) Systematic biases in panel surveys due to differential non-response. *Journal of the American Statistical Association*, 65, 1338–1349.

Wright, T. (2014) *A Simple Method of Exact Optimal Sampling Allocation under Stratification with Any Mixed Constraint Patterns*. Research Report Series (Statistics # 2014–07). U.S. Bureau of Census, Washington DC.

Wright, T. (2016) *Two Optimal Exact Sample Allocation Algorithms: Sampling Variance Decomposition Is Key*. Research Report Series (Statistics # 2016–03). U.S. Bureau of Census, Washington DC.

Wu, C.F.J., and Hamada, M. (2002) *Experiments: Planning, Analysis and Parameter Design Optimization*. Wiley, New York.

Xie, M., Tan, K.C., and Goh, K.H. (1998) Fault tree reduction for reliability analysis and improvement. *Frontiers in Reliability, Series on Quality, Reliability and Engineering Statistics*, 4, 411–428.

Yates, F. (1949) *Sampling for Censuses and Surveys*. Griffin, London.

Young, F.W., and Hamer, R.M. (1994) *Theory and Applications of Multidimensional Scaling*. Erlbaum, Mahwah, NJ.

Zacks, S., Rogatko, A., and Babb. J. (1998) Optimal Bayesian-feasible dose escalation for cancer phase I trails. *Statistics & Probability Letters*, 38, 215–220.

Zenga, M. (2007) Inequality curve and inequality index based on the ratios between lower and upper arithmetic means. *Statistics and Applications*, 5(1), 3–27.

Zenga, M.M., Radaelli, P., and Zenga, M. (2012) Decomposition of Zenga's inequality index by sources. *Statistics and Applications*, 10(1), 3–21.

Index